Neural Computing: An Introduction

Neural Computing: An Introduction

R Beale and T Jackson

Department of Computer Science, University of York

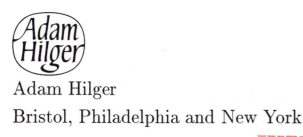

Adam Hilger

Bristol, Philadelphia and New York

British Library Cataloguing in Publication Data

Beale, R.
    Neural computing : an introduction
    1. Artificial intelligence
    I. Title   II. Jackson, T.
    008.3

ISBN 0-85274-262-2

Library of Congress Cataloging-in-Publication Data are available

Section 3.5 is based on material from *Perceptrons* by M Minsky and S Papert, pp 164–9, ©1969 MIT Press.

Published under the Adam Hilger imprint by IOP Publishing Ltd
Techno House, Redcliffe Way, Bristol BS1 6NX, England
335 East 45th Street, New York, NY 10017-3483, USA
US Editorial Office: 1411 Walnut Street, Suite 200, Philadelphia, PA 19102

Printed in Great Britain by J W Arrowsmith Ltd, Bristol

To our parents

Contents

Preface

Neural computing is one of the most rapidly expanding areas of current research, attracting people from a wide variety of disciplines. These people all bring a different background to the area, and one of the aims of this book is to provide a common ground from which new developments can grow. Another aim is to explain the basic concepts of neural computation to an interested audience, and so this book is about the whole field of neural networks, covering all the major approaches and their important results; more especially, it is an introduction, developing the concepts and ideas from their simple basics through their formulation into powerful computational systems.

We have tried to assume as little as possible in the reader, and have attempted to set the book out in a clear and logical order. As well as showing the basic concepts behind each of the major approaches, these have been put in the context of their historical development, so that the reasoning behind the model is as apparent as its basic features. We have explained the ideas in English first, and have also included the mathematical treatments of the models, for which we make no apology. The subject is a mathematical one, and the precise formulations of mathematics demonstrate things that are difficult to explain in English alone. The important derivations and proofs are also included since they form a vital part of the development of the area as a whole, and indicate points at which apparently insurmountable problems were reached, and then overcome. As well as the mathematics, we have included the basic algorithms for the major approaches. These algorithms are a series of steps that implement the ideas behind each of the different models, and much can be gained if they are converted into code for a computer, since familiarity with the concepts is best gained by "hands-on" experience. They are also useful to those without access to computers, however, since they set out the steps required in a straightforward way, and may clarify some of the comments in the surrounding text.

Various little pictures crop up again and again in the book—these icons contain information about the content of the section in which they appear, and are described below.

This represents a section of text that is mathematical in nature. Not all the mathematical parts of the book are indicated like this, since they are usually important to the overall understanding; however, sections with this icon at the start can be skipped over at first reading without losing too much of the flow of discussion, if you are not familiar with the mathematics. They do contain material that is useful and interesting, however, and we would suggest that an effort should be made to look at them on a second reading.

This represents an algorithm for a particular model, and is designed to help you locate them within the book. Ideally, they should be looked at when they are encountered in the text, but again they can be omitted on first reading.

This symbol appears at the end of every chapter, where we have tried to compress the major concepts into a succinct summary. These represent the bare bones of the subject, and can be used to check that you have followed the main features of the chapter.

The very end of each chapter contains some suggestions for further reading. We have tried to avoid referencing lots of papers from many different sources, and instead have directed attention towards specific books that deal with a particular subject in depth. Detailed references can then be gleaned from these, if appropriate.

What's Where

Chapter 1: Introduction contains the background to the subject; the first section takes a light philosophical look at the differences between humans and computers. The second section describes a simplified model of the real brain. Analogies are drawn between artificial neural nets and their biological counterparts.

Chapter 2: Pattern Recognition contains a discussion of the basic concepts and ideas of pattern recognition, necessary since the majority of tasks that are required of neural networks involve recognition. It gives an overview of the predominant standard approaches to the area, so that the place and function of neural systems can be understood clearly.

Chapter 3: Basic Neuron develops the basic model of the single-layer perceptron, its learning rule, operation and features, and its partitioning of pattern space. It shows the problems associated with classifying the exclusive-or (XOR) and other non-linear problems.

Chapter 4: Multilayer Perceptrons develops the model from Chapter 3, showing how it can be altered to make it more powerful. It covers the concepts of back-propagation, including the generalised delta rule, gradient descent, and the concepts of feature extraction by hidden units. The energy landscape is evoked to give a visualisation of energy minimisation and the problems of local minima. The chapter also contains a section on the applications of the method to real problems.

Chapter 5: Kohonen Self-Organising Networks looks at a different paradigm, that of unsupervised learning. It looks at the formation of self-organising topological maps and contains a detailed description of one of the most influential applications of neural technology, that of the phonetic typewriter.

Chapter 6: Hopfield Networks contains the description of the fully connected Hopfield net, and its probabilistic partner the Boltzmann machine, as well as a look at some analogies with physical systems, and optimisation problems.

Chapter 7: ART revolves around an explanation of the more biologically-inspired approach of Grossberg, highlighting the differences between the architecture and approach of this system to those covered earlier.

Chapter 8: Associative Memory expands on the current techniques for implementing associative memories and associative neural networks, including the RAM nets of Aleksander, the matrix memories of Willshaw, and the ADAM system. The parallels between associative memories and other neural networks are explored.

Chapter 9: Into the Looking Glass views the future of neural computing, and gives an insight into some of the exciting recent developments that point the way forward.

How to read this book

If the aim of the reader is to properly understand neural networks, we would suggest that the book is read in order. An alternative approach for those particularly interested in the more recent developments in the field, and who have some background knowledge, is to briefly read Chapter 3 to familiarise yourself with the foundations of the subject, followed by Chapters 4 through to 7. For those with an interest in the biological implications, Chapter 1 should be read first, followed by Chapter 7, then Chapters 2–6 and 8 will add context.

Without Whom ...

Most projects involve the collaboration of a number of people, and writing this book has been no exception. We are greatly indebted to a number of people who, through their support, comments and criticism, have kept us enthused, put up with our moanings, and helped us transform the initial idea into reality. We appreciate their assistance and effort, especially that of our colleagues at the University of York, and in particular those within our research group, who have contributed freely to discussions and made us question our most basic assumptions—these people have made our work that much more interesting and life that much more fun. In particular, we would like to thank Dr. Jim Austin for his support throughout, and for the academic arguments that have aided our understanding and abetted our enthusiasm. Chris Higgins explained the depths of LaTeX, and wrote the macros—the good bits of the typesetting are due to him, whilst the bad bits are all our own work. Personal thanks must extend to Julia, for her encouragement, support and late-night typing. To Derek Wills, for dragging us into the world of neural networks in the first place. And to Janet, who has tolerated preoccupations, shared disappointments, read many drafts, and still found nice things to say. The book is better because of them all—if it is bad, the blame lies with us.

Russell Beale and Tom Jackson.

1

Introduction

1.1 HUMANS AND COMPUTERS

Human beings are more intelligent than computers. Why is that
said? Has it anything to do with the fact that I am human, and I
don't want to think that a lump of silicon and metal can do all I can
do? Or is it because computers are different from us, in terms of the
operations they perform? For instance, calculating the sum of a few
hundred eight and nine digit numbers is a trivial calculation for a
computer, but it is asking a lot of even the most adept person. Does
that make the computer more intelligent than us? An initial answer
to that may be that it does—so consider crosswords instead. Some
of us are excellent at doing crosswords, others are terrible—but we
all can usually manage a couple of clues in an easy one. This sort of
task is immensely difficult for a computer, however. Solving cross-
word puzzles usually involves working out what an obscurely worded
clue is referring to, and takes what we term leaps of intuition and
guesswork, where we follow lines of enquiry that are not immediately
obvious but are sparked off by some recollection or idea that happens to come along. Computers can't do this at all well—perhaps we
would adjust our definition and say that they were logical, and could
only do logical things well. Then we may consider vision; an activity
that would appear perfectly logical to us—look at something, work
out what it was, and give it a name, and possibly do something with
it; if it were a cup of tea, we'd recognise it as such, and drink it—if
the object was a football coming fast towards us, we'd have to decide on some more complex course of action. But again, computers
are very bad at performing simple visual tasks. They have a job to

1

distinguish simple items, and as for actually controlling an arm to pick it up or something similar, that requires exceptionally complex techniques.

Perhaps the problem is, because computers can do some of the tasks that we do in a fraction of the time, such as add numbers, or recall names and addresses accurately months after it first knows about them, we expect them to be like us in many other ways as well. We are then disappointed when they do not perform as well as we want them. This problem is really the one that people in artificial intelligence want to tackle, but their efforts, even after 30 years of high-quality research, are not sufficient to allow them to make the claim that they have computer systems that are artificially intelligent in any general sense that we would recognise. The aim of artificial intelligence could be summed up as trying to make computers behave as they do in the movies—there, the computers seem to always work, and are evidently superior to the humans that run round them; a far cry from real life and the unpaid wages or huge bill that arise because the computer has "done something wrong".

Why then can't computers do the things that we do? One of the answers would appear to be in the nature of their design. We would not unreasonably expect that things that are designed to operate in similar ways to exhibit similar behaviour. If we look inside a computer, we see a number of chips, containing miniature circuits and components, plugged into a circuit board with resistors and other things on. If we look inside the brain, we see nothing like such an ordered structure: our initial inspection reveals nothing more than a convoluted mass of homogeneous grey matter. Further investigation reveals that it too contains components, but these are all arranged in an immensely complex fashion, each connected to thousands of others. Perhaps it is this difference in design that can account for the difference in performance between the systems. Computers are designed to carry out one instruction after another, extremely rapidly, whereas our brains work with many more slower units. Whereas a computer can typically carry out a few million operations every second, the units in the brain respond about ten times per second. However, they work on many different things at once, which computers can't do. The computer is a high-speed, serial machine, and is used as such, compared to the slow, highly parallel nature of the brain. Given this, is it so surprising that the computer fails to perform in the same way as the brain? It manages tasks which suit its design very well: counting is an essentially serial activity, as is adding, with one thing done after another, and so the computer can beat the brain any time. For vision or speech recognition, the problem is a highly parallel one, with many different and conflicting inputs triggering many different and conflicting ideas and memories, and it is only the combining of all these different factors that allow us to perform such feats—but then, our brains are able to operate in parallel easily and so we leave the computer far behind. Perhaps the lesson here is that one thing may be good for one purpose but not necessarily for another: just because my computer can add up numbers, should I expect it to solve vision problems easily?

The conclusion that we can reach from all of this is that the problems that we are trying to solve are immensely parallel ones. They require the processing of lots of different items of information which all interact to provide a solution. The knowledge required to solve these problems comes from many different sources, each with its own contribution to make to the final output. The brain, with its parallel design, is able to represent and store this knowledge in an accessible way. It is also able to process this knowledge along with the many different stimuli that it receives, due again to the parallel nature of its operation. Speed is not the important factor—parallelism is, and the brain is ideally suited to the task.

The approach of neural computing is to capture the guiding principles that underly the brain's solution to these problems and apply them to computer systems. We do not know how the brain represents high-level information, so cannot mimic that, but we do know that it uses many slow units that are highly interconnected. In modelling the brain's basic systems, we should end up with a solution that is intrinsically suited to parallel problems rather than serial ones. These parallel models should be able to represent knowledge in a parallel fashion, and process it in a similar way. We can *simulate* these structures in a serial fashion, though, so we do not need to build new computers. However, the inherently parallel nature of artificial neural network systems does make them amenable to implementation on parallel machines, which may offer advantages in terms of speed and ultimate reliability; after all, that is how the brain has done it. To rework an old adage—we want the right architecture for the right job.

In the following chapters we look at how the study of real neural systems has allowed us to model the parallelism that exists in the brain, and has given us artificial neural networks that have behaviour that is heading toward what we really want. Whilst we are copying the parallelism of the brain, it would also seem sensible to notice other useful features of real neural systems and see if we can incorporate them into our new networks.

Perhaps one of the most important of the features is that the brain is able learn things—it can teach itself. Learning from example is the way in which as children we pick up speech, learn to write, eat and drink, and develop our own set of standards and morals. The same is not true of conventional computer systems. In these, the computer usually has a long and complicated program, which gives it specific instructions as to what to do at every stage in its operation. Each step of the way has to be spelled out, and it is fairly obvious that we don't work this way at all, since when we come to write such programs, it takes us many hours of patient and careful work to write down exactly what we mean in a form that the computer can understand. For large programs, these instructions may be many millions of lines long, and one mistake can cause all sorts of unexpected things to happen; such mistakes are known as bugs, and are the blight of a computer scientist's life. Indeed, these mistakes are recognised as being immensely difficult to avoid, and most large programs have many bugs in them. If you were to buy a new car, you would not expect it to go wrong, but if you were to buy a new piece of software, you would be extremely surprised if it worked without a mistake. Bugs are accepted as a fact of life. But wouldn't it be nice if instead of having to develop a program to do a task, you could simply let the computer observe the task for a while, so that it could learn by example? And who knows, it may find a better way of doing it than you, so that it was more efficient that a simple program would be. It would probably have bugs in it initially, so that it occasionally did something wrong—but it would learn from its mistakes and not repeat the error.

1.2   THE STRUCTURE OF THE BRAIN

The human brain is one of the most complicated things that we have studied in detail, and is, on the whole, poorly understood. We do not have satisfactory answers to the most fundamental of questions such as "what is my mind?" and "how do I think?". Nevertheless, we do have a basic understanding of the operation of the brain at a low level. It contains approximately ten thousand million (10^{10}) basic units, called neurons. Each of these neurons is connected to about ten thousand (10^4) others. To put this in perspective, imagine an Olympic-sized swimming pool, empty. The number of raindrops that it would take to fill the pool is approximately 10^{10}. You'd also need at least a dozen full address books if you were to be able to contact 10^4 other people.

The neuron is the basic unit of the brain, and is a stand-alone analogue logical processing unit. The neurons form two main types, local processing *interneuron* cells that have their input and output connections over about 100 microns, and output cells that connect different regions of the brain to each other, connect the brain to muscle, or connect from sensory organs into the brain. The operation of the neuron is a complicated and not fully understood process on a microscopic level, although the basic details are relatively clear. The neuron accepts many inputs, which are all added up in some fashion. If enough active inputs are received at once, then the neuron will be activated and "fire"; if not, then the neuron will remain in its inactive, quiet state. A representation of the basic features of a neuron is shown in figure 1.1.

The *soma* is the body of the neuron. Attached to the soma are long, irregularly shaped filaments, called *dendrites*. These nerve processes are often less than a micron in diameter, and have complex branching shapes. Their intricate shape resembles that of a tree in

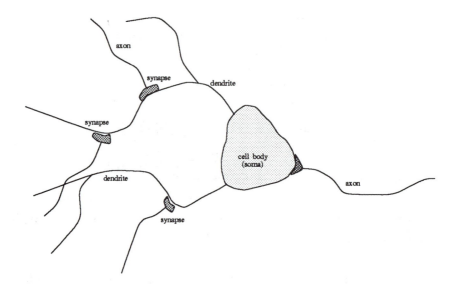

Figure 1.1 The basic features of a biological neuron.

winter, without leaves, whose branches fork and fork again into finer structure. The dendrites act as the connections through which all the inputs to the neuron arrive. These cells are able to perform more complex functions than simple addition on the inputs they receive, but considering a simple summation is a reasonable approximation.

Another type of nerve process attached to the soma is called an *axon*. This is electrically active, unlike the dendrite, and serves as the output channel of the neuron. Axons always appear on output cells, but are often absent from interneurons, which have both inputs and outputs on dendrites. The axon is a non-linear threshold device, producing a voltage pulse, called an *action potential*, that lasts about 1 millisecond (10^{-3}s) when the resting potential within the soma rises above a certain critical threshold. This action potential is in fact a series of rapid voltage spikes. See figure 1.2 for an illustration of this "all-or-nothing" principle.

The axon terminates in a specialised contact called a *synapse* that couples the axon with the dendrite of another cell. There is no direct linkage across the junction; rather, it is a temporary chemical

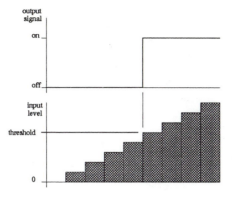

Figure 1.2 The input to the neuron body must exceed a certain threshold before the cell will fire.

one. The synapse releases chemicals called *neurotransmitters* when its potential is raised sufficiently by the action potential. It may take the arrival of more than one action potential before the synapse is triggered. The neurotransmitters that are released by the synapse diffuse across the gap, and chemically activate gates on the dendrites, which, when open, allow charged ions to flow. It is this flow of ions that alters the dendritic potential, and provides a voltage pulse on the dendrite, which is then conducted along into the next neuron body. Each dendrite may have many synapses acting on it, allowing massive interconnectivity to be achieved. At the synaptic junction, the number of gates opened on the dendrite depends on the number of neurotransmitters released. It also appears that some synapses excite the dendrite they affect, whilst others serve to inhibit it. This corresponds to altering the local potential of the dendrite in a positive or negative direction. A single neuron will have many synaptic inputs on its dendrites, and may have many synaptic outputs connecting it to other cells.

1.2.1   Learning in Biological Systems

Learning is thought to occur when modifications are made to the effective coupling between one cell and another, at the synaptic junction. Figure 1.3 shows the important features of the synapse in more detail. The mechanism for achieving this seems to be to facilitate

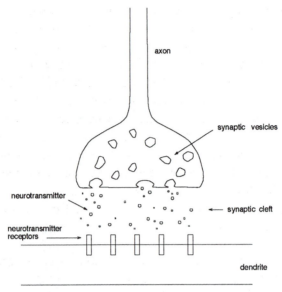

Figure 1.3 The synapse. Neurotransmitters released from the synaptic vesicles diffuse across the synaptic cleft and trigger the receivers on the dendrite.

the release of more neurotransmitters. This has the effect of opening more gates on the dendrite on the post-synaptic side of the junction, and so increasing the coupling effect of the two cells. The adjustment of coupling so as to favourably reinforce good connections is an important feature of artificial neural net models, as is the effective coupling, or *weighting*, that occurs on connections into a neuronal cell.

1.2.2   The Organisation of the Brain

The brain is organised into different regions, each responsible for different functions, and in humans this organisation is very marked. The largest parts of the brain are the cerebral hemispheres, which occupy most of the interior of the skull. They are layered structures, the most complex being the outer layer, known as the *cerebral cortex*, where the nerve cells are extremely densely packed to allow great interconnectivity. Its function is not fully understood, but we can get some indication of its purpose from studies of animals that have had it removed. A dog, for example, can still move in a coordinated manner, will eat and sleep, and even bark if it is disturbed. However, it also becomes blind and loses its sense of smell—more significantly, perhaps, it loses all interest in its environment, not responding to people or to its name, nor to other dogs, even of the opposite sex. It also loses all ability to learn. In effect, it loses the characteristics that we generally refer to as indicating intelligence—awareness, interest and interaction with an environment, and an ability to adapt and learn. Thus the cerebral cortex seems to be the seat of the higher order functions of the brain, and the core of intelligence.

The cerebral cortex has been the subject of investigation by researchers for many years, and is slowly revealing its secrets. It demonstrates a localisation of functions, in that different areas of the cortex fulfill different functions, such as motion control, hearing, and vision. The visual part of the cortex is especially interesting. In the visual cortex, electrical stimulation of the cells can produce the sensation of light, and detailed analysis has shown that specific layers of neurons are sensitive to particular orientations of input stimuli, so that one layer responds maximally to horizontal lines, whilst another responds to vertical ones. Although much of this structure is genetically pre-determined, the orientation-specific layout of the cells appears to be learnt at an early stage. Animals brought up in an environment of purely horizontal lines do not develop neuron structures that respond to vertical orientations, showing that these structures are developed due to environmental input and not purely from genetic pre-determination. This *self-organisation* of the visual cortex, so called since there is no external teacher to guide the development of these structures, is discussed further in Chapter 5, where the work of Kohonen has shown that such feature maps can be developed in artificial neural systems as a consequence of simple learning rules.

1.3   LEARNING IN MACHINES

The ability to learn is not unique to the biological world, and is captured within our neural network models. However, the concept of machine learning goes against many of the commonly held beliefs about computers; that they can do only what they are programmed to do, and cannot adapt to their surroundings. Whilst it is true on an atomic level that the program controls the machine, the behaviour that results does not have to be so rigid and deterministic as is commonly felt. Having a computer learn to respond correctly to a given input, or learn to play a game, is not a simple concept, and it is often felt that complicated programs and systems are required to achieve behaviour such as this, that many would class as one of the requirements for intelligence. The purpose of this section is to discuss these beliefs with reference to a machine called MENACE, developed by Donald Michie in the early 1960's, which learns how to play the game of noughts and crosses. What is interesting is that MENACE requires no expensive hardware or clever programming; it is constructed from matchboxes, each containing a number of beads.

MENACE (Matchbox Educable Noughts And Crosses Engine) consists of 288 matchboxes, one for every possible distinct board position that the opening player can encounter. Each matchbox is then filled with a random selection of coloured beads, each colour representing a move to a corresponding colour on the board. The game is played by selecting at random a bead from the matchbox that corresponds to the current board position, with the colour selected determining the machine's move. The first game is played, with the machine moving completely at random. When the game is over, the outcome is fed back into the machine so that it can adapt its behaviour in the light of the outcome; i.e. it can learn to play better next time. This is achieved by reinforcing all the moves that were ultimately successful, when the machine won, and by decreasing the chance of it making the same bad moves that led to defeat. Learning therefore occurs by adding a bead of the same colour to boxes representing a successful series of moves, or by removing a bead of the colour that led to defeat. A draw means that the number of beads remains the same. This slow process of learning from experience continues, until the probability of the machine making a good move far outweighs the chance of it making a bad one. Once it has learnt, the machine is then almost invincible, and the best that can be hoped for is to consistently draw with it.

This simple device demonstrates some important features of machine learning. It usually takes some time for a machine to achieve a good probabilistic solution to a problem, which is what MENACE achieves, but it is possible, given that the reinforcement learning takes place. This reinforcement learning is analogous to that which is thought to occur in the brain when the efficacies of the synaptic junction are increased in order to promote the recurrence of a neural event. No external teacher is required to train MENACE with the tactics of noughts and crosses; it learns the most successful strategies purely by example, when the final result is used to modify the machine's subsequent performance. MENACE also has no specific location in which the information needed to play successfully is stored; rather, it is distributed throughout the machine in the probabilities of coloured beads in each box. It treats the process of learning to play the game as a series of smaller sub-problems; each box corresponds to a single situation and a number of possible moves, not enough on its own to play the game. Learning occurs in each of these boxes, and each box is unaware of the state of the other boxes that participate in the game—it only knows the outcome. Successful learning can occur since the behaviour of the system as a whole is stochastic, and increasing the chance of good moves from one box increases the probability of an eventual win.

But perhaps the most surprising feature of MENACE is that a pile of matchboxes can learn to play a game of noughts and crosses at all.

1.4   THE DIFFERENCES

We have seen that the brain is excellent at performing many of the tasks that we would like computers to perform, such as vision, speech recognition, learning by example and so on. We have also seen that the brain is structured in such a way as to make the accomplishment of these tasks as easy as possible, which inevitably means that there are certain things on which it cannot perform so well. The compromises that have evolved have been dictated by the most important functions, where the ability to learn and adapt, to see and interpret sounds has been more important than the ability to add up a series of numbers accurately. The brain manages to accomplish these complex tasks with an apparent minimum of effort due to its highly developed structure, that of a massively parallel system, in which many simple processing elements share the job of working out what is going on, rather than trying to make one fast node do all the work. This division of labour has other advantages as well; since many neurons are involved at any one time, the contribution made by a single one is not too important, and so if it happens to go wrong, it is unlikely to affect the others in a significant way. This distribution of work, known as *distributed processing*, therefore has the advantage that it is tolerant of errors here and there. Indeed, because the brain can learn, it is able to adjust to the permanent loss of one of its neurons and can bring in new ones. This ability to function with only some of the processing elements working correctly is known in computing circles as *fault tolerance*, for the obvious reason that a system, such as the brain, can tolerate faults within it without producing nonsense as output. This is a vital feature of the operation of the brain, since every day a few neurons die as part of the natural course of events. More are lost if the brain gets bumped about, but it continues working as if nothing had happened. In cases of continuing damage, parallel distributed systems exhibit what is known as *graceful degradation* where the performance of the system slowly falls from a high level to a reduced level, but without dropping catastrophically to zero. Compare this to the situation of a single unit working hard to calculate lots of things quickly enough to reach a correct output—if this element breaks down, then there is no hope of obtaining a sensible answer, and no hope of coping with the situation by transferring some of the work elsewhere. There is nowhere else for it to go; a classic example of putting all your eggs into one basket!

Computers are very different in structure, however. Rather than being comprised of many millions of relatively slow, highly interconnected processing elements like the brain, they consist of one (or occasionally, on modern machines, maybe two or a few more) exceptionally fast processor, which is capable of many million calculations per second—this makes it good at performing simple, repetitive actions like adding numbers, but poorer at the task of processing the vast quantities of different types of data that a vision system requires. They also suffer from not embracing the distributed approach in areas apart from speed, in that they are intrinsically intolerant of faults. If the processor in a computer breaks, that's it: the screen may go blank; worse, an aircraft may crash, or all the lights go out in a city—the consequences may be far-reaching and difficult to correct or even anticipate.

These problems have led to the current interest in developing computer systems that adopt the principles developed by millions of years of evolution—that is, keep it simple, keep it joined up, and have lots of it to share the load.

 Summary

- Brain is parallel, distributed processing system.
- Basic processing unit called the neuron.
- Approximately 10^{10} neurons each connected to 10^4 others.
- Operation of neuron: fires pulse down axon when sufficient input received from dendrites. Connections via chemical junctions called synapses.
- Learning increases efficacy of synaptic junction.
- Machines can learn through positive reinforcement.
- Cerebral cortex shows local areas of specialised function.

FURTHER READING

The following references are to books and articles that address the whole field of neural network research, and they constitute a good starting point for those readers wishing to follow up references to a particular aspect of the area.

1. *Parallel Distributed Processing*, Volumes 1, 2, and 3. J. L. McClelland & D. E. Rumelhart. Volume 1 covers the foundations and many of the current approaches and models, whilst volume 2 looks at the subject from a more biological viewpoint. Volume 3 contains a tutorial and software.

2. An Introduction to Computing with Neural Nets. Richard P. Lippmann. In *IEEE ASSP Magazine*, April 1987. An excellent, concise overview of the whole area.

3. An Introduction to Neural Computing. Teuvo Kohonen. In *Neural Networks*, volume 1, number 1, 1988. A general review.

4. *Neurocomputing: Foundations of Research*. Edited by Anderson and Rosenfeld. MIT Press, 1988. An expensive book, but excellent for reference, it is a collection of reprints of most of the major papers in the field.

5. *Neural Computing: Theory and Practice*. Philip D. Wasserman. Routledge, Chapman & Hall, 1989. An introductory text. Well-written.

There are many journals in which papers on neural computing appear, but the following list should provide a basis for further research.

1. *Neural Networks*. Published bi-monthly.

2. *Network: Computation in Neural Systems*. Published quarterly.

3. *Neural Information Processing Systems (NIPS)*. Annual conference proceedings.

4. *IJCNN Conference*. Annual conference proceedings. Used to be the IEEE conference.

2

Pattern Recognition

2.1 INTRODUCTION

Pattern recognition—a strange heading for Chapter 2 of a book on neural computing. At least it is, until we point out that pattern recognition (in one form or another) is currently the dominating area for the application of neural networks. It is a large area of computer science in itself, and those wishing to pursue neural networks will not get far before bumping into some of the issues raised by the task of pattern recognition. The material that we will discuss in this chapter, namely a definition of pattern recognition and an overview of current techniques, is essential background reading. Much of the mathematics overlaps with that of neural networks, and, to a large extent, the two areas are tackling the same problems. It will only be the briefest of introductions to pattern recognition techniques, but we hope to cover all the basic issues that will affect our understanding of neural networks.

2.2 PATTERN RECOGNITION IN PERSPECTIVE

To appreciate what the pattern recognition problem is all about let us consider a task that is fairly basic to the majority of people—reading. A significant proportion of the information that we absorb (i.e. that is applied to our biological "neural networks") is presented to us in the form of patterns. The text that you are reading now is presenting you with complex and varied patterns in the form of strings of letters. Before we even start to consider the far reaching cognitive issues of language processing, the visual system must first solve the pattern recognition problem. That is, recognising the neatly aligned ink stains on this page as alphabetic characters!

The fact that our visual system copes with this task effortlessly we naturally take for granted. However, if we present this task to a computer, we soon begin to realise the enormous complexity of the problem. This "classification" is one of the simpler pattern recognition tasks. It could be resolved using a template matching technique where each letter is read into a fixed size frame and the frame compared to a template of all the possible characters. This is the solution used in simpler applications, for example matching parts on a factory production line, where we can predict the variety of shapes that are likely to be encountered. Consider however what would happen if we encountered a change in the typeface of the text in our reading task. Unless we had a second template set for the new font the technique would probably fail miserably at the classification task.

And further consider the case for handwritten text—it would prove a near impossibility to provide templates to cope with the widely varying patterns in cursive script (students may well appreciate the problem of attempting to decipher lecturers' blackboard notes!). Text processing is just one example of the pattern recognition problem. The difficulties described above are further complicated when we turn our attention to processing images, speech or even stock market trends.

Later chapters will describe how neural networks provide computational techniques that are able to deal with these problems. First though it is necessary to provide a more formal definition of pattern recognition techniques.

2.3   PATTERN RECOGNITION—A DEFINITION

The fundamental objective for pattern recognition is classification: given an input of some form can we analyse that input to provide a meaningful *categorisation* of its data content?

A pattern recognition system can be considered as a two stage device. The first stage is feature extraction. The second is classification.

We define a feature as a measurement taken on the input pattern that is to be classified. Typically, we are looking for features that will provide a definite characteristic of that input type. For example, thinking about our text processing problem again, if we wish to distinguish the letter 'F' from the letter 'E' we would need to compare the number of vertical and horizontal strokes in the character. Feature extraction is rarely as trivial as the example we have given and often poses the greater part of the recognition problem.

The classifier is supplied with the list of measured features. Its task is to map these input features onto a classification state, that is, given the input features, the classifier must decide which type of class category they match most closely. Classifiers typically rely on distance metrics and probability theory to do this. Before we look at these techniques however, we would first like to provide some useful definitions.

2.4   FEATURE VECTORS AND FEATURE SPACE

Classification is rarely performed using a single measurement, or feature, from the input pattern. Usually, several measurements are required to be able to adequately distinguish inputs that belong to different categories (or classes, as they are normally called). If we make n measurements on our input pattern, each of which is a unique feature, then we can use algebraic notation to create a set of these features and call it a *feature vector*. The dimensionality of the vector, that is, the number of elements in it, creates an n dimensional *feature space*.

The simplest way to describe feature space is to consider a simple two-dimensional example—that is we will make two measurements on the pattern to form the feature vector. A rather trivial example might be distinguishing ballet dancers from rugby players (as if it isn't obvious!). Thinking about the problem, we might decide that two distinctive measurements that categorise each type are height and weight. If we make a series of height and weight measurements on typical examples of each, then we can plot the range of readings in a two-dimensional Euclidean plane (x_1, x_2) that defines our feature space, as shown in figure 2.1.

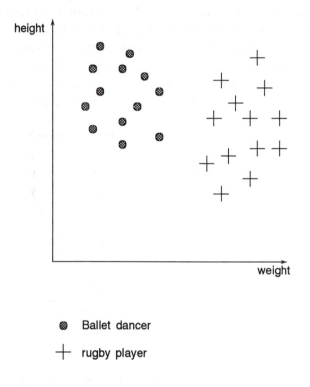

Figure 2.1 A two-dimensional Euclidean feature space.

This plot of our measurements helps us to visualise the concept of our feature space. It does, of course, get a little tricky trying to visualise anything above a dimension of three.

2.5   DISCRIMINANT FUNCTIONS

Discriminant functions are the basis for the majority of pattern recognition techniques. Let us think again about our two-dimensional rugby-player/ballet-dancer classification problem shown in figure 2.1. Looking at the spread of the measured samples we can see they form two distinct clusters.

The classifier stage is required to assign a class to these clusters, and also assign a new input example to one of the classes. Looking at the spread of the data in the clusters, we could intuitively decide that some line drawn between the two classes could arbitrarily separate them. If we could define such a dividing boundary for our data, classification would become a process of deciding on which side of the boundary any new input falls, as shown in figure 2.2.

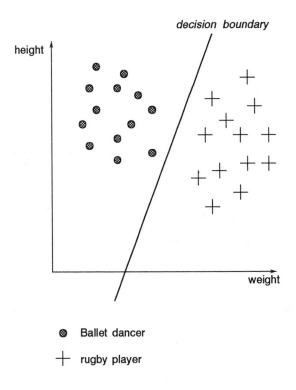

Figure 2.2 A linear classification decision boundary.

The mathematical definition of such a decision boundary is a "discriminating function". It is a function that maps our input features onto a classification space—in the example above, by defining a plane that would separate the two clusters. The above example is an over-simplification of the problem and rarely would our decision boundary be so easily defined. Even in this simplistic example however it can be appreciated that there are an infinite number of boundaries we could have drawn to separate the two regions. In practice, though, it is advisable to make the discriminant function as simple as possible (we have to compute the function at some stage, so the simpler the better!).

In the case above it is fairly obvious that the simplest function that would separate the two clusters is a straight line. This represents a very widely used category of classifiers known as linear classifiers.

2.6   CLASSIFICATION TECHNIQUES

Pattern classification techniques fall into two broad categories—numeric and non-numeric. Numeric techniques include deterministic and statistical measures which can be considered as measures made on the geometric pattern space. Non-numeric techniques are those which take us into the domain of symbolic processing that is dealt with by such methods as fuzzy sets. For the purposes of this book we shall only consider the numeric techniques as they have far more bearing on our discussion of neural computing. That is not to say that people do not use neural networks for symbolic data manipulation (in the traditional artificial intelligence sense)—many research groups are in fact putting a great deal of effort into this concept. However, it is perhaps a little esoteric to be included in an introductory text, so barring a brief discussion in the final chapter on future trends in neural networks we shall restrict ourselves to numeric methods.

We have already touched on deterministic methods in our discussion of discriminant functions. We shall be looking more closely at a particular implementation of discriminant function analysis known as "K nearest neighbour" as well as taking a further look at linear classification. For the statistical approach we shall discuss Bayesian classification which uses probabilistic estimation of class membership. These choices have been made on the grounds that they are very widely used classification techniques, so widespread in fact that many applications of neural networks are ultimately benchmarked against them for performance. For this reason, if no other, it will be very useful to familiarise yourself with them.

2.6.1   Nearest Neighbour Classification

Consider the diagram of figure 2.3.

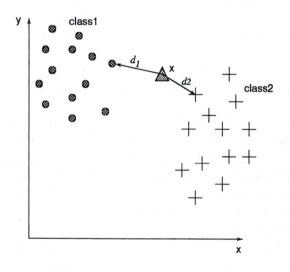

Figure 2.3 Classification by comparison to the "nearest neighbour".

We have two classes represented in pattern space and we wish to decide to which of the two the unclassified pattern, X, belongs. Nearest neighbour techniques, in essence, make a decision based on the shortest distance to the neighbouring class samples—they assign it to whichever class it appears to be closest to (not an unreasonable assumption). Formally, that defines a discriminant function $f(X)$ by:

$$f(X) = \text{closest(class1)} - \text{closest(class2)}$$

For class patterns that are well separated in pattern space, as we have shown in figure 2.1, this technique will work by assigning $f(X)$ negative to, say, class 1 membership and $f(X)$ positive to class 2 membership. The range of problems that this simple dichotomiser may be applied to is, however, rather restricted (at least in terms of useful performance). Consider the case of a rogue pattern, figure 2.4, that has class membership of one class but does in fact lie closer to another class—it is not typical of its class type but is included none the less. In this instance, if our unclassified input is measured against the rogue sample, it will invariably result in misclassification. The solution to this fairly basic problem is to take several distance measures against many class samples such that the effect of any rogue measurement made is likely to be averaged out. This is "K" nearest neighbour classification—where "K" is the number of neighbouring samples against which we decide to measure.

2.6.2   Distance Metrics

Nearest neighbour methods pose the problem of finding a reliable way of measuring the distance from one class sample to another. Obviously, we need to specify a distance metric that will allow us to measure the similarity of pattern samples in the geometric pattern space. In practice, several methods are used.

• Hamming distance measure.

The most basic measure, and one that is widely used because of its simplicity, is the Hamming distance measure. For two vectors

$$
\begin{aligned}
X &= (x_1, x_2, \ldots) \\
Y &= (y_1, y_2, \ldots)
\end{aligned}
$$

the Hamming distance is found by evaluating the difference between each component of one vector with the corresponding component

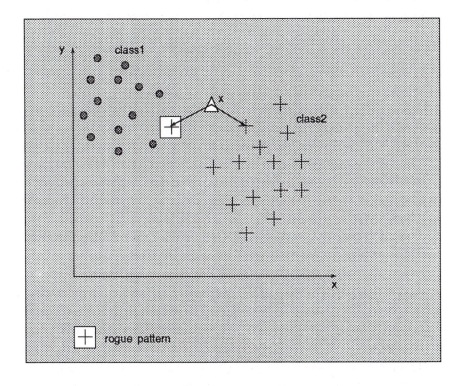

Figure 2.4 Measuring to the nearest neighbour can produce errors in classification if a rogue sample is selected.

of the other, and summing these differences to provide an absolute value for the variation between the two vectors. The measure is defined by:

$$H = \sum(|x_i - y_i|)$$

The Hamming distance is often used to compare binary vectors. It is perhaps obvious that in this case the Hamming distance provides a value for the number of bits that are different between two vectors. In actual fact the Hamming distance measure for binary data can be performed simply by the exclusive-OR function since

$$|x_i - y_i| \text{ is equivalent to } x_i \text{XOR } y_i$$

- Euclidean distance measure.

One of the most common metrics used is the Euclidean Distance measure. Consider an example in a rectangular coordinate system where we have two vectors (X and Y) that we wish to find the distance between them ($d(X,Y)$).

The shortest distance, shown dotted on figure 2.5, is the Euclidean distance which is defined by:

$$d(X,Y)_{\text{euc}} = \sqrt{\left(\sum_{i=1}^{n}(X_i - Y_i)^2\right)}$$

where n is the dimensionality of the vector.

For the two-dimensional example we have drawn, this gives us:

$$d(X,Y)_{\text{euc}} = \sqrt{(x_1 - y_1)^2 + (x_2 - y_2)^2}$$

There is nothing too strange about that, of course, as it is simply Pythagoras's theorem for the sides of a triangle. A special case is given for binary vectors where the metric is then equivalent to the square root of the Hamming distance.

The Euclidean metric is widely used mainly because it is simple to calculate. For binary input vectors the metric reduces to a special case which is mathematically equivalent to the *square root* of the Hamming distance. The metric is used in a neural network learning algorithm discussed in Chapter 5.

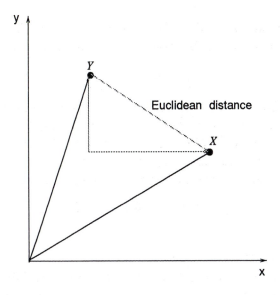

Figure 2.5 The Euclidean distance measure.

- City block distance (Manhattan)

A simplified version of the Euclidean distance measure is the city block measure. This method performs the Euclidean measure without calculating the squared or square root functions. Thus

$$D_{cb} = \sum_{n} |X_j - Y_j|$$

The effect of this, apart from the obvious one that it is much faster to compute than the Euclidean, is that points of equal distance from a vector lie on a square boundary about the vector, as opposed to a circular boundary for the Euclidean. This is illustrated in figure 2.6.

The enclosing circle shown is the Euclidean boundary for equidistant points about the vector. For the city block distance, anything falling on the square boundary will yield the same distance value. As you no doubt realise, this does introduce some error into the measure, but this is accepted as a compromise between accuracy and speed of calculation.

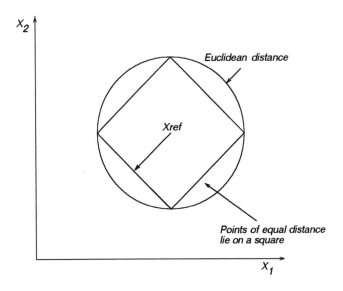

Figure 2.6 City block distance metric.

- Square distance.

Simplifying the Euclidean distance measure still further—but consequently adding still more error—we have the square distance, shown in figure 2.7. With this measure the distance between two vectors is defined as the maximum of the differences between each element of the two:

$$D_{sq} = \text{MAX}|X_i - Y_i|$$

This again defines a square boundary for points equidistant from a vector. It is however a larger square than that of the city block, and is consequently a coarser measure. As before, however, the error is tolerated as a compromise between speed and accuracy.

That concludes a brief look at distance metrics; it is by no means exhaustive but we hope that it it least indicates the possible techniques available for comparing the similarity of two vectors. In the following section, we focus again on the idea of discriminating functions using decision boundaries rather than comparison methods.

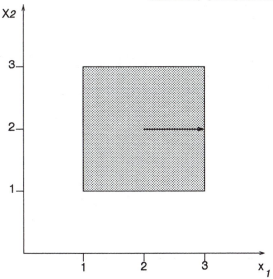

Figure 2.7 The square distance metric.

2.7   LINEAR CLASSIFIERS

Linear classification is a pattern recognition technique that is encountered time and time again in the field of neural networks. We shall provide an overview of a linear classifier, describe how it can be used in pattern recognition, and will endeavour to unravel the mysteries of the *non-linearly separable* problem that has plagued neural network research since the late 1960's.

In the preceding discussion about partitioning the pattern space by discriminant functions, we have already paved the way for this discussion of linear classifiers. Let us think again about the simple two-dimensional, two-class discrimination problem, illustrated in figure 2.2. We wish to classify an input into one of two possible classes, A or B. We have already described how the classes may be separated in pattern space by the use of a linear decision boundary, but how can we implement such a decision boundary in the case of real pattern data, and how is the position of the separating boundary chosen?

In figure 2.8 we show our pattern space with a new vector added. This vector we will describe as a *weight vector*, W, and its orientation in pattern space will be used to define a linear decision boundary.

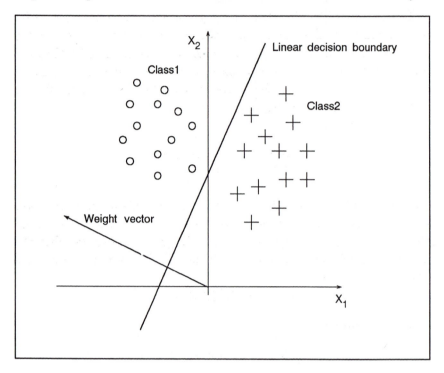

Figure 2.8 Discriminating classes with a linear decision boundary. Note the inclusion of the weight vector.

The decision boundary defines a discriminating function $f(X)$ of the form:

$$f(X) = \sum_{i=1}^{n} W_i X_i$$

where

$$X_i \;\; = \;\; i-\text{th component of an input vector}$$

$$W_i \;=\; i-\text{th component of a weight vector}$$
$$N \;=\; \text{dimensionality of the input vector.}$$

The output of the function for any input will be either a *positive* or *negative* value depending upon the the value of the weight vector and the input vector. If we let a positive output indicate that the input vector belongs to, say, class A and a negative output indicate class B then we have a decision mechanism that simply looks for the sign of $f(X)$ for any input value.

Class definition:

$$\text{if } f(X) \;>\; 0 \;=\; \text{class A}$$
$$\text{if } f(X) \;<\; 0 \;=\; \text{class B}$$

The problem lies in actually finding a suitable weight vector that will give these results for all inputs from class A and class B. If we expand the discriminant function using matrix algebra we can visualise the dependence of the output on the value of the weight vector. We have:

$$f(x) = \sum W_i X_i - \theta$$

This expands to:

$$f(x) = (|W|.|X| \cos \phi) - \theta$$

where ϕ is the angle between the vector X and W.

The $\cos \phi$ term swings between $+/- 1$, consequently any value of ϕ greater than $+/- 90$ degrees between the weight vector and the input will reverse the sign of the output of $f(X)$. This is clearly a *straight line* decision boundary since the crossover point is at $+90$ or -90 degrees. We can see that the function does indeed give us a decision boundary but we are no closer to realising the position of this boundary or finding the correct components for the weight vector.

There are two parameters that control the position of the decision boundary in the pattern space—these are the slope of the line and the y-axis intercept (standard geometry of a straight line). The slope of the line in the function is actually determined by the value of the weight vector. We can see this if we consider the crossover point, or boundary condition, when the output of the classifier is zero.

We have:

$$\sum W_i X_i - \theta = 0$$
$$X_1 \times W_1 + X_2 \times W_2 - \theta = 0$$

Rearranging this gives us:

$$X_2 = -W_1/W_2 \times X_1 + \theta/W_2$$

Comparing this to the equation of a straight line ($y = mx + c$) we can see that the slope of the line is controlled by the ratio of the weight values W_1 and W_2 and the intercept is controlled by the bias value, θ.

Thus far we have proved that if we have the correct value for the weight vector we can indeed perform the discriminating process and set the position of the decision boundary. What we have not shown yet is the critical part—namely finding the weight vector. This, unfortunately, is a not a trivial problem! It is most usually found by iterative trial and error methods that modify the weight values according to some *error function*. The error function typically compares the output of the classifier with a desired response and gives an indication of the difference between the two. If we considered a general logic implementation of the discriminant function we can start to appreciate the scale of the problem. For an n-bit binary input there will be 2^n possible input patterns. Classifying these using $+/-$ dichotomy means that there are 2^{2n} possible logic functions that would map the n inputs to the correct output value. The linear classifier, however, can only perform a small number of these possible mappings—those that are in fact defined as *linearly separable*. Linear separability is a subject that has strong links with the potted history of neural network research, and it will be discussed in length in Chapter 3. For now, we shall define linear separable problems as those that can be satisfied using a single hyperplane decision surface.

The examples we have discussed so far only show linear classifiers discriminating between two possible classes. However, linear classifiers can also be used to separate more than two classes, by arranging many decision boundaries and performing several tests to satisfy the conditions for each class. As an example, in a four-class problem (A, B, C, D), the decision boundaries can be selected to test between A or BCD, if the result is not A then test for B or CD, if not B then test for C or D. Similarly for difficult class boundary conditions the decision surface can be split up in a piecewise fashion, as shown in figure 2.9.

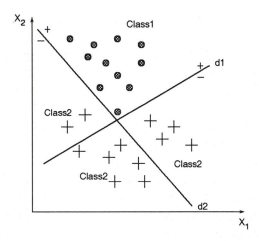

classification	sign of decision line	
	d1	d2
Class1	+	+
Class2	+	−
	−	+

Figure 2.9 Piecewise linear classification for a non-linearly separable pattern.

In non-linearly separable problems it is also possible to introduce the required non-linearity into the decision surface by applying a non-linear transformation to the data before it is passed to the classifier stage. This technique is described as a Φ machine and such preprocessing of pattern data before passing it to a pattern classifier is common practice. A transform is found that will map the patterns into a new coding that is capable of being classified using a linear classifier. The major drawback of this approach is that it can be slow.

2.7.1   Conclusion

This concludes our look at deterministic methods for pattern classification. It is far from complete, but hopefully it will provide enough background information to put the forthcoming discussions of neural computing techniques into perspective.

2.8   STATISTICAL TECHNIQUES

Statistical techniques play a major part in pattern classification. Without launching into a deep statistical treatment (you will be glad to hear) we wish to discuss the concept of Bayesian classification. It is an important analytical technique, and is very powerful and widely used. Using techniques of this kind also has the added advantage of forcing us to think harder about the statistical nature of the data that we are dealing with in pattern recognition problems. Any method that makes us think long and hard about the nature of the problem with which we are dealing—particularly about the characteristics of the data—cannot be too highly valued. We will make the point early in the book that applying any of the techniques described in this book, with any degree of success, relies heavily on one understanding the nature of the problem in the first place. That may seem like a fairly obvious statement to make but in the light of recent claims for the "magical" problem solving abilities of neural networks we feel it is perhaps a necessary one. Addressing our problem statistically we can gain a very useful insight into the nature of the pattern data that we are dealing with—as well as perhaps a more intuitive feel for what makes pattern recognition problems often so difficult to solve.

Bayesian classification relies on the basic statistical theory of probabilities and conditional probabilities. For pattern classification we are using measurements taken from patterns (i.e. the components of our feature vector) to make an estimate of the likelihood, or probability, of a pattern belonging to a particular class. Let us give some basic definitions; if we let G_i, $i = 1, 2, \ldots, n$ be our list of possible groups or classes then we can define the probability of a pattern belonging to a class as $P(G_i)$ (where $0 \leq P(G_i) \leq 1$). Using conditional probabilities allows us to include knowledge we already have about the pattern to to improve our estimate of class membership. For example, if we try to predict the possibility of an ace being dealt from a pack of cards after, say, ten cards have been dealt out—then it is easier to make that prediction if we know which ten cards have already been dealt. If they are dealt face up, and we have already seen four aces dealt from the pack, then we could state—without too much reason for doubt—that the eleventh card dealt will not be an ace. A trivial example, perhaps, but it illustrates that including prior knowledge into our estimates will have a considerable influence on their reliability.

Given two events, X and Y, we can define conditional probability as the probability of event X given the *occurrence* of event Y. This is written as $P(X|Y)$. For pattern recognition, the prior knowledge that we are combining with the estimate of class membership comprises the data measurements taken from the pattern—that is our feature vector $X = (x_1, x_2, x_3, \ldots, x_n)$. Our classification problem can now be stated as: given a set of measurements, X, what is the likelihood, or probability, of it belonging to a class G_i— i.e. $P(G_i|X)$.

This is where Bayes's rule enters—it is a formalisation of the statement that we have already made. If we make measurements on a pattern to give us a feature vector, X, on a pattern that we know must come from one class of $G_1, G_2, \ldots, G_n$ then Bayes's rule assigns it to a class on the following basis.

Decide x belongs to class i for

$$P(G_i|X) > P(G_j|X) \quad \text{for } i = 1, 2, \ldots, n \quad i \neq j$$

Put simply, it says that we assign a pattern to the class that has the highest conditional probability of the vector X belonging to it. It may come as something of a surprise to find that it can be proven that this will provide us with the best estimate that we could hope for—if we measure our performance in terms of smallest average error rate.

In practice, however, it's not quite so simple (it sounded too good to be true, didn't it?). The difficulties arise in actually defining the conditional probabilities required for Bayes's rule. More often than not they are in fact not known and must be estimated by some means. Obviously the accuracy of the estimates will ultimately determine the performance of the classifier in these circumstances. How then, are they estimated? Typically, this involves making assumptions about the pattern data and describing unknown distributions in the data with "models". The problem can be simplified if we rearrange the constraints of the conditional probability and ask the question, given that we know the pattern must belong to one of n groups, what is the probability of obtaining that pattern vector in each of the possible groups. We denote this $P(X|G_i)$. Although we do not know the absolute value of this probability, we can in fact approximate it by using a model probability distribution and assuming that it will follow the same trend. We may not seem to have gained a great deal by this step, but there is in fact a simple relationship between $P(G_i|X)$ and $P(X|G_i)$, that is known as Bayes's law:

$$P(G_i|X) = \frac{P(X|G_i).P(G_i)}{\sum_j P(X|G_j)P(G_j)}$$

We defined $P(G_i)$ earlier as the probability of a pattern belonging to the class G_i—this can be found without too much difficulty. In most practical situations $P(X|G_i)$ is estimated by assuming that it follows the "normal" distribution. Although it may appear that this is a somewhat arbitrary decision this model does in fact have many useful properties that make it a particularly apt choice. The most obvious is the fact that it is a distribution that does occur in many situations—or at least a close approximation of it. It is also a good approximation to many other distributions. Its most endearing quality is the fact that it is easy to work with—its distribution has been well researched and there is a large pool of knowledge from which to draw when using it.

By adopting models for the conditional probability $P(X|G_i)$ based on a standard distribution—the normal curve—and applying Bayes's law, we can define a relatively straightforward statistical classifier. The performance will depend on how close the pattern data does actually fit the model selected, but generally Bayesian classifiers can be optimised to perform extremely well.

Bayesian classifiers also have further merits that justify their widespread use. They can, in fact, be made to look like a linear classifier by making some simple assumptions about the pattern data. What is more, this can be done in such a way that we finish with a *deterministic* process that apparently makes no reference to statistics at all. The simplifications or assumptions that we have to make about the pattern data relate to the spread of the normal distributions of the classes.

If we revert to a view of the distributions in two-dimensional coordinates for a two class problem, it can be shown that the pattern space is most effectively partitioned by a quadratic decision surface. Whilst being relatively easy to use it can in fact be modified to the simplest case of a linear classifier. This is achieved by making the assumption that both the class distributions have equal covariance matrices. This amounts to saying that the distributions both have the same overall shape and spread. The consequence of this is that the most accurate partition of the pattern space is in fact achieved by a linear surface. This is demonstrated in figure 2.10 with a straight line separating a simple two-class case.

The proof that Bayes's law does in fact reduce to a linear classifier can be performed analytically by solving Bayes's law for the boundary conditions of the two classes. The solution reduces to a function that is of the form $y = mx + c$, that is, that of a straight line. Those who are sufficiently versed in statistical theory may wish to pursue this proof in the pattern recognition texts referenced in the bibliography, but we shall do no more here than quote the result.

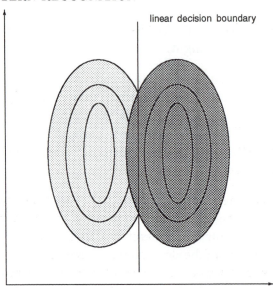

Figure 2.10 Bayesian classification reduces to linear classification under certain conditions.

2.9   PATTERN RECOGNITION—A SUMMARY

This chapter has been written expressly for those who are coming to neural networks with no background knowledge of pattern recognition. The methods we have discussed will at least provide the bare essentials that will be drawn upon in the following chapters on neural computing. Neural computing is a subject that spans many diverse fields of science—none of which is more fundamental to a solid grasp of the area than an appreciation of the classification methods used in pattern recognition. The methods that we have described in this chapter are the ones that will be most often referred to in neural networks. We hope that they will leave you adequately "armed" to appreciate the strengths and weaknesses of neural computing that are discussed in the rest of the book.

 ## Summary

- Pattern recognition—feature extraction and classification.
- Features are pattern measurements used for comparison.
- Discriminant functions partition up feature space.
- A number of different distance metrics are used.
- Linear classification occurs when classes can be separated by a single linear decision boundary. Classes that cannot be separated this way are termed non-linearly separable.

FURTHER READING

1. *Pattern Recognition.* M. James. BSP Professional Books (Oxford), 1987. A good basic introduction to computer based pattern recognition techniques.

2. *Adaptive, Learning and Pattern Recognition Systems.* J. M. Mendel, K. S. Fu. Academic Press (New York and London), 1970. An old book but a very complete treatment of classical pattern recognition theory and adaptive systems.

3. *Self Organisation and Associative Memory*, third edition. T. Kohonen. Springer-Verlag, 1990. Chapter 2; A tutorial discussion of pattern mathematics—a brief but useful revision of matrix algebra techniques.

4. *Adaptive Pattern Recognition & Neural Networks.* Y. H. Pao. Addison Wesley, 1989. A good discussion of pattern recognition concepts in a neural network context.

3

The Basic Neuron

3.1 INTRODUCTION

In Chapter 1, we have examined the structure of the brain, and
found it to be a highly developed mechanism that is relatively poorly
understood, but capable of immensely impressive tasks. We have
seen that many of the things that we would like computers to be
able to do, the brain manages exceptionally well, and the idea behind
neural computing is that by modelling the major features of the brain
and its operation, we can produce computers that exhibit many of
the useful properties of the brain.

We have noted the complexity of the structure of the brain; however, it can be viewed as a highly interconnected network of relatively
simple processing elements. We need a model that can capture the
important features of real neural systems in order that it will exhibit similar behaviour. However, the model must deliberately ignore many small effects, if it is to be simple enough to implement
and understand. This extraction of a few features deemed important
and disregard of all others is a general characteristic of modelling;
the aim of a model is to produce a simplified version of a system
which retains the same general behaviour, so that the system can be
more easily understood.

3.2 MODELLING THE SINGLE NEURON

We will firstly consider the features of a single neuron and how we
can model it. The basic function of a biological neuron is to add

up its inputs, and to produce an output if this sum is greater than some value, known as the *threshold* value. The inputs to the neuron arrive along the dendrites, which are connected to the outputs from other neurons by specialised junctions called synapses. These junctions alter the effectiveness with which the signal is transmitted; some synapses are good junctions, and pass a large signal across, whilst others are very poor, and allow very little through. The cell body receives all these inputs, and fires if the total input exceeds the threshold value. This simple biological neuron is shown in figure 3.1.

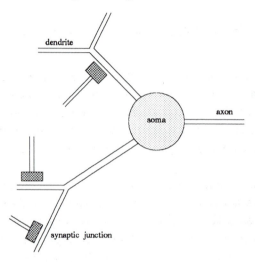

Figure 3.1 The basic features of a biological neuron.

Our model of the neuron must capture these important features. We can summarise them as follows:

- The output from a neuron is either on or off.
- The output depends only on the inputs. A certain number must be on at any one time in order to make the neuron fire.

The efficiency of the synapses at coupling the incoming signal into the cell body can be modelled by having a multiplicative factor on each of the inputs to the neuron. A more efficient synapse, which transmits more of the signal, has a correspondingly larger weight, whilst a weak synapse has a small weight.

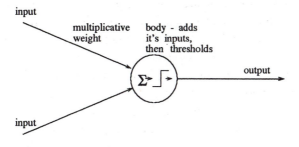

Figure 3.2 Outline of the basic model.

So now we have our basic model of the neuron, shown in figure 3.2. It performs a weighted sum of its inputs, compares this to some internal threshold level, and turns on only if this level is exceeded. If not, it stays off. Because the inputs are passed through the model neuron to produce the output, the system is known as a *feedforward* one.

We need to formulate this mathematically. If there are n inputs, then there are n associated weights on the input lines. The model neuron calculates the weighted sum of its inputs; it takes the first input, multiplies it by the weight on that input line, then does the same for the next input, and so on, adding them all up at the end. This can be written as

$$
\begin{aligned}
\text{total input} \quad &= \quad \text{weight on line 1} \times \text{input on 1} + \\
&\qquad \text{weight on line 2} \times \text{input on 2} + \cdots + \\
&\qquad \text{weight on line } n \times \text{input on } n \\
&= \quad w_1 x_1 + w_2 x_2 + w_3 x_3 + w_4 x_4 + \cdots + w_n x_n \\
&= \quad \sum_{i=1}^{n} w_i x_i
\end{aligned}
$$

This sum then has to be compared to a certain value in the neuron, the threshold value. This thresholding process is accomplished by comparison; if the sum is greater than the threshold value, then output a 1, if less, output a 0. This can be seen graphically in figure 3.3 where the x-axis represents the input, and the y-axis the output.

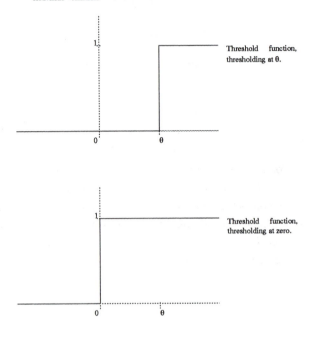

The thresholding function is alternatively known as the "step" function, or the "Heaviside" function.

Threshold function, thresholding at θ.

Threshold function, thresholding at zero.

Figure 3.3 The thresholding function.

Equivalently, the threshold value can be subtracted from the weighted sum, and the resulting value compared to zero; if the result is positive, then output a 1, else output a 0. This is also shown in figure 3.3; notice that the shape of the function is the same, but now the jump occurs at zero. The threshold effectively adds an offset to the weighted sum. An alternative way of achieving the same effect is to take the threshold out of the body of the model neuron and connect it to an extra input value that is fixed to be "on" all the time. In this case, rather than subtracting the threshold value from the weighted sum, the extra input of +1 is multiplied by a weight equal to minus the threshold value, $-\theta$, and added in as well as all the other inputs—this is known as *biasing* the neuron. The value of $-\theta$ is therefore known as the neuron's *bias* or *offset*. Both approaches are equivalent, and either is acceptable.

Calling the output y, we can write

$$y = f_h \left[\sum_{i=1}^{n} w_i x_i - \theta \right]$$

where f_h is a step function (actually known as the *Heaviside* function) and

$$f_h(x) = 1 \qquad x > 0$$
$$f_h(x) = 0 \qquad x \leq 0$$

so that it does what we want. Note that the function produces only a 1 or a 0, so that the neuron is either on or off.

If we use the approach of biasing the neuron, we can define an extra input, input 0, which is always set to be on, with a weight that represents the bias applied to the neuron. The equation describing the output can then be written as

$$y = f_h \left[\sum_{i=0}^{n} w_i x_i \right]$$

Notice that the lower limit of the summation has changed from 1 to 0, and that the value of the input x_0 is always set to 1.

This model of the neuron, shown in figure 3.4, was proposed in 1943 by McCulloch and Pitts. Their model came about in much the same way as we have developed ours, and stemmed from their research into the behaviour of the neurons in the brain. It is important to look at the features of this McCulloch-Pitts neuron. It is a simple enough unit, thresholding a weighted sum of its inputs to get an output. It specifically does not take any account of the complex patterns and timings of actual nervous activity in real neural systems, nor does it have any of the complicated features found in the body of biological neurons. This ensures its status as a *model*, and not a *copy*, of a real neuron, and makes it possible to implement on a digital computer. This is the strength of the model—now we need

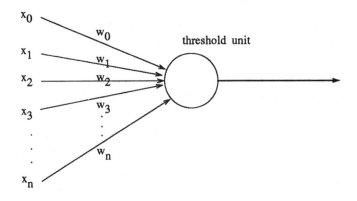

Figure 3.4 Details of the basic model.

to investigate what can be achieved using this simple design. The arrangement of the connections between the neurons is important, but, continuing our trend of choosing simple models to get an idea of what is happening in a complicated real-world situation, we shall for the time being consider only one layer of neurons, where we study the outputs of the neurons under a known set of inputs.

The model neurons, connected up in a simple fashion, were given the name "perceptrons" by Frank Rosenblatt in 1962. He pioneered the simulation of neural networks on digital computers, as well as their formal analysis. In his book *"Principles of Neurodynamics"*, he describes these perceptrons as simplified networks in which certain properties of real nervous systems are exaggerated whilst others are ignored. He stated that they are not intended to serve as detailed copies of any real nervous system; in other words, he realised at this early stage that he was dealing with a basic model. This fact is often lost in the popular press as the idea of computer "brains", based on these techniques, grabs the imagination. We are not attempting to build computer brains, nor are we trying to mimic parts of real brains—rather we are aiming to discover the properties of models that take their behaviour from extremely simplified versions of natural neural systems, usually on a massively reduced scale as well. Whereas the brain has at least 10^{10} neurons, each connected to 10^4

others, we are concerned here with maybe a few hundred neurons at most, connected to a few thousand input lines.

3.3   LEARNING IN SIMPLE NEURONS

We need a mechanism for achieving learning in our model neuron. Connecting these neurons together may well produce networks that can do something, but we need to be able to *train* them in order for them to do anything useful. As we have seen before, it is the ability of these networks to learn that makes them especially useful. We also want to find the simplest learning rule that we can, in order to keep our model understandable. As is often the case in neural computing, inspiration comes from looking at real neural systems.

Young children are praised for doing well in a maths test. They are scolded for rushing across the road without looking. Dogs are given titbits to encourage them to come when called. In general, good behaviour is reinforced, whilst bad behaviour is reprimanded. We can transfer this idea to our network. We must try to reinforce behaviour that we want repeated and discourage things that we do not. If we have two groups of objects, for example one group of several differently written A's, and the other of B's, we may want our neuron to tell the A's from the B's, as in figure 3.5. We want it to output a 1 when an A is presented and a 0 when it sees a B.

We need to think about our model neuron, and examine its behaviour, to see how we can include the concept of learning within our simple design. The guiding principle is to allow the neuron to learn from its mistakes. If it produces an incorrect output, we want to reduce the chances of that happening again; if it comes up with correct output, then we need do nothing. If we set up the neuron with *random* weights on its input lines, corresponding to a starting state in which it knows *nothing*, we can present an A. The neuron will perform the weighted sum of the inputs, and compare this to the threshold. If it exceeds the threshold, it will output a 1, whilst if it doesn't, it will output a 0. The likelihood that it will get it correct are 50:50 at first, since the inputs to the neuron have only a random chance of exceeding the threshold. Let us assume it does

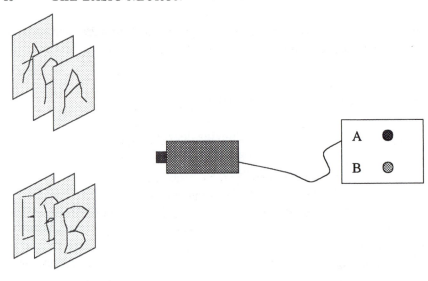

Figure 3.5 Can we tell the A's from the B's?

get the correct answer, then we do not need to do anything, since the model has been successful. But if the neuron produces a 0 when we show it an A, we want to increase the weighted sum so that next time it will exceed the threshold and so produces the correct output, a 1. We would do this by increasing the weights. So, to reinforce the chances of getting a 1, we want to increase the weights.

For inputs that are B's, we want the neuron to produce 0's. This means that we want the weighted sum of the inputs to be less than the threshold, and so each time we present a B we want to decrease the weights, to try and force the neuron to produce a zero next time.

This means that for the network to learn, we want to increase the weights on the active inputs when we want the output to be active, and to decrease them when we want the output to be inactive. We can achieve this by adding the input values to the weights when we want the output to be on, and subtracting the input values from the weights when we want the output to be off. This defines our learning rule. Notice that only those inputs which are active at the time will be affected; this is sensible since the inactive ones do not contribute to the weighted sum, and so changing them will not affect the result for the particular input in question, but may well upset what has already been learnt.

This learning rule is a variant on that proposed in 1949 by Donald Hebb, and is therefore called Hebbian learning. Hebb postulated his rule, that of reinforcing *active* connections only, from his studies of real neuronal systems. The slightly modified version that we use retains the notion of only affecting active connections, but we have allowed them to be strengthened or weakened. We can do this because we can see which way to alter the weights as we know what the result should be. Since the learning is guided by knowing what we want to achieve, it is known as *supervised learning*. We have developed these ideas of learning from the point of view of the model and common sense, and have derived a learning rule that is not unlike the one postulated for biological systems. It is the dominant method used today in learning models.

This simple idea for learning actually remained untested until 1951, when Marvin Minsky and Dean Edmonds built a "neural network"—it was quite a machine! This large-scale device used 300 tubes, lots of motors and clutches, and a gyropilot from a World War II bomber to move its 40 control knobs. The position of these knobs represented the memory of the machine, and Minsky and Edmonds spent a long time watching the machine at play, as it adjusted the knobs and moved several things all at once. The huge amount of wiring connecting it up was full of poorly soldered joints and incorrect connections, but the random nature of the whole system allowed it to continue working even when some of the tube "neurons" broke down as well. This mechanical contraption was probably the first realisation of a learning network.

Our learning paradigm can be summarised as follows:

- set the weights and thresholds randomly
- present an input
- calculate the actual output by taking the thresholded value of the weighted sum of the inputs
- alter the weights to reinforce correct decisions and discourage incorrect decisions—i.e. *reduce* the error

- present the next input etc.

3.3.1   The perceptron learning algorithm

The learning procedure that we have described can be written as the following algorithm, which can be used to implement a perceptron network on a computer by coding the steps in any programming language.

Perceptron Learning Algorithm

---

1. Initialise weights and threshold
Define $w_i(t), (0 \leq i \leq n)$, to be the weight from input i at time t, and θ to be the threshold value in the output node. Set w_0 to be $-\theta$, the bias, and x_0 to be always 1.

Set $w_i(0)$ to small random values, thus initialising all the weights and the threshold.

2. Present input and desired output
Present input $x_0, x_1, x_2, \ldots, x_n$ and desired output $d(t)$

3. Calculate actual output

$$y(t) = f_h \left[\sum_{i=0}^{n} w_i(t)x_i(t) \right]$$

4. Adapt weights

if correct	$w_i(t+1)$	$=$	$w_i(t)$
if output 0, should be 1 (class A)	$w_i(t+1)$	$=$	$w_i(t) + x_i(t)$
if output 1, should be 0 (class B)	$w_i(t+1)$	$=$	$w_i(t) - x_i(t)$

Note that weights are unchanged if the net makes the correct decision. Also, weights are not adjusted on input lines which do not contribute to the incorrect response, since each weight is adjusted by the value of the input on that line, x_i, which would be zero.

This is the basic perceptron algorithm. However, various modifications have been suggested to this basic algorithm. The first is to introduce a multiplicative factor of less than one into the weight adaption term. This has the effect of slowing down the change in the weights, making the network take smaller steps towards the solution. This alteration to the algorithm entails replacing step 4 with the following:

4. Adapt weights—modified version

$$
\begin{array}{rrcl}
& \text{if correct} & w_i(t+1) & = & w_i(t) \\
\text{if output 0, should be 1 (class A)} & & w_i(t+1) & = & w_i(t) + \eta x_i(t) \\
\text{if output 1, should be 0 (class B)} & & w_i(t+1) & = & w_i(t) - \eta x_i(t)
\end{array}
$$

where $0 \leq \eta \leq 1$, a positive gain term that controls the adaption rate.

Another algorithm of a similar nature was suggested by Widrow and Hoff. They realised that it would be best to change the weights by a lot when the weighted sum is a long way from the desired value, whilst altering them only slightly when the weighted sum is close to that required to give the correct solution. They proposed a learning rule known as the Widrow-Hoff delta rule, which calculates the difference between the weighted sum and the required output, and calls that the *error*. Weight adjustment is then carried out in proportion to that error. This means that during the learning process, the output from the unit is *not* passed through the step function—however, actual classification is effected by using the step function to produce the +1 or 0 indication as before.

The error term Δ can be written

$$\Delta = d(t) - y(t)$$

where $d(t)$ is the desired response of the system, and $y(t)$ is the actual response. This takes care of the addition or subtraction, since if the desired output is 1 and the actual output is 0, $\Delta = +1$ and so the weights are increased. Conversely, if the desired output is 0 and the actual output is $+1$, $\Delta = -1$ and so the weights will be decreased. Note that weights are unchanged if the net makes the correct decision, since $d(t) - y(t) = 0$.

The learning algorithm is basically the same as for the basic perceptron, except this time step 4 is replaced by

4. Adapt weights—Widrow-Hoff delta rule

$$\Delta = d(t) - y(t)$$
$$w_i(t+1) = w_i(t) + \eta \Delta x_i(t)$$
$$d(t) = \begin{cases} +1, & \text{if input from class A} \\ 0, & \text{if input from class B} \end{cases}$$

where $0 \leq \eta \leq 1$, a positive gain function that controls the adaption rate

Neuron units using this learning algorithm were called ADALINEs (adaptive linear neurons) by Widrow, who also connected many of them together into a many-ADALINE structure, or MADALINE.

Another alternative proposed is to use inputs that are not 0 or 1 (binary), but are instead -1 or $+1$, known as *bipolar*. Using binary inputs means that input lines with 0's on them are not trained, whereas bipolar values allow all the inputs to be trained each time. This simple alteration helps to speed up the convergence process, but often leads to confusion in the literature as some authors discuss binary inputs and others bipolar ones. Effectively, they are equivalent, and the use of one or the other is usually a matter of personal preference.

3.4 THE PERCEPTRON: A VECTORIAL PERSPECTIVE

If we write the inputs to a perceptron as a vector $\mathbf{X} = (x_0, x_1, \ldots, x_n)$ we can think about the algorithm in a vectorial fashion. This vector $\mathbf{X}$ has n elements, and so is called n-dimensional. We can only really imagine at most three dimensions, but it is still possible to get a feel for what is going on. If we write the weights as another vector $\mathbf{W} = (w_0, w_1, \ldots, w_n)$ then we can replace the weighted sum with the identical vector dot product, ie.

$$\sum_{i=0}^{n} w_i x_i \equiv \mathbf{W} \cdot \mathbf{X}$$

The learning algorithm for the perceptron ensures that the weights are adapted to reduce the error each time. We can understand how the perceptron learning procedure works on an intuitive level by examining the behaviour of the weight vector as the perceptron learns patterns. If we continue our example consisting of patterns of A's and B's, we can see that they can be represented in pattern space as shown in figure 3.6.

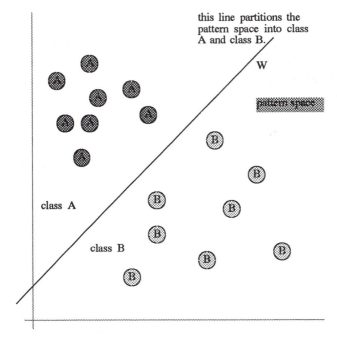

Figure 3.6 Two distinct sets of patterns drawn in 2-d pattern space.

The solution to classifying these patterns is to produce a dividing line between them, such as the line W in the diagram. Points above the line can be regarded as representing patterns from class A, whilst those below the line are from class B. This line is what we want our perceptron to discover for itself. A line such as this, which separates two classes in pattern space, is said to *partition* the space into two classes.

The perceptron generates this line by adjusting the values of the elements of the weight vector, as prescribed by the learning procedure, so that inputs from the top side of the line produce a 1 as output, and inputs from below the line produce a 0. The perceptron starts with a random weight vector (see step 1 of the learning procedure) that points anywhere in the pattern space. A pattern is presented, and the learning procedure ensures that if the output is incorrect, the weight vector is altered to reduce the error. This is achieved by moving the vector a finite amount towards the ideal weight vector. Eventually, the weight vector becomes the ideal weight vector, and gives no error for inputs from either class, thus partitioning the pattern space successfully. The perceptron has then "learnt" to distinguish between A's and B's. The behaviour of the weight vector can be visualised with the help of figure 3.7. The effect

Figure 3.7 Behaviour of the weight vector in pattern space.

of the learning process on the line that partitions the pattern space is shown in figure 3.8. As learning progresses, the partitioning of the classes evolves from the initial random state into a correct one.

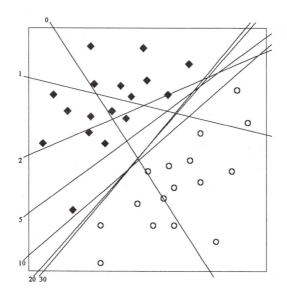

Figure 3.8 Evolution of the classification line from an initial, random orientation into one that successfully classifies the two classes.

3.5   THE PERCEPTRON LEARNING RULE: PROOF

We have seen, intuitively, how the perceptron learning rule produces a solution; in this section we *prove* this fact. This proof was first proposed by Rosenblatt. His influential result stated that, given it is possible to classify a series of inputs, then a *perceptron network will find that classification*. In other words, he proved that the perceptron weight vector would eventually align itself with the ideal weight vector, and would not oscillate around it for ever. The proof relies on vector notation, and contains some mathematics. It follows the approach taken by Minsky and Papert in their book

Perceptrons; it can be skipped on first reading, but none of the concepts are too difficult, and the mathematics describes what is going on in a succinct and elegant way.

Definitions:

The input patterns are assumed to come from a space which has two classes, $\mathbf{F}^+$ and $\mathbf{F}^-$. We want the perceptron to respond with $+1$ if the input comes from $\mathbf{F}^+$ and -1 if it comes from $\mathbf{F}^-$.

Consider the set of input values x_i as a vector in i-dimensional space, called $\mathbf{X}$, and the set of weights w_i as another vector in the same space, denoted by $\mathbf{W}$. To make things simple, let us assume that the vectors $\mathbf{X}$ are of unit length—it makes no difference to the final result, except clarifying the maths a bit.

Increasing the weights is performed by adding $\mathbf{X}$ to $\mathbf{W}$ vectorially, and decreasing them means subtracting $\mathbf{X}$ from $\mathbf{W}$.

Replacing $\sum_{i=0}^{n-1} w_i x_i(t)$ by the vector notation $\mathbf{W} \cdot \mathbf{X}$ produces the following algorithm.

START :
    Choose any value for $\mathbf{W}$
TEST :
    Choose an $\mathbf{X}$ from $\mathbf{F}^+ \cup \mathbf{F}^-$

$$\text{If } \mathbf{X} \in \mathbf{F}^+ \text{ and } \mathbf{W} \cdot \mathbf{X} > 0 \text{ goto TEST}.$$
$$\text{If } \mathbf{X} \in \mathbf{F}^+ \text{ and } \mathbf{W} \cdot \mathbf{X} \leq 0 \text{ goto ADD}.$$
$$\text{If } \mathbf{X} \in \mathbf{F}^- \text{ and } \mathbf{W} \cdot \mathbf{X} < 0 \text{ goto TEST}.$$
$$\text{If } \mathbf{X} \in \mathbf{F}^- \text{ and } \mathbf{W} \cdot \mathbf{X} \geq 0 \text{ goto SUBTRACT}.$$

ADD :
    Replace $\mathbf{W}$ by $\mathbf{W} + \mathbf{X}$
    Goto TEST .
SUBTRACT :
    Replace $\mathbf{W}$ by $\mathbf{W} - \mathbf{X}$
    Goto TEST .

Notice that we go to SUBTRACT when $\mathbf{X}$ is from class $\mathbf{F}^-$, and if we consider that going to SUBTRACT is the same as going to ADD but with $\mathbf{X}$ replaced by $-\mathbf{X}$, then we can rewrite the procedure as follows.

START:

Choose any value for $\mathbf{W}$

TEST:

Choose a $\mathbf{X}$ from $\mathbf{F}^+ \cup \mathbf{F}^-$

If $\mathbf{X} \in \mathbf{F}^-$	change the sign of	$\mathbf{X}$
If $\mathbf{W} \cdot \mathbf{X} > 0$	goto	TEST
	otherwise goto	ADD .

ADD:

Replace $\mathbf{W}$ by $\mathbf{W} + \mathbf{X}$

Goto TEST .

We can simplify the algorithm still further, if we define $\mathbf{F}$ to be

$\mathbf{F}^+ \cup -\mathbf{F}^-$ i.e., $\mathbf{F}^+$ and the negatives of $\mathbf{F}^-$, we can say

---

START:

Choose any value for $\mathbf{W}$

TEST:

Choose any $\mathbf{X}$ from $\mathbf{F}$

If $\mathbf{W} \cdot \mathbf{X} > 0$	goto	TEST
	otherwise goto	ADD .

ADD:

Replace $\mathbf{W}$ by $\mathbf{W} + \mathbf{X}$

Goto TEST .

The *convergence theorem* then states that the program will only go to ADD a finite number of times. This is what we have to prove.

Proof: Assume there is a unit vector $\mathbf{W}^*$, which partitions up the space, and a small positive fixed number δ such that

$$\mathbf{W}^* \cdot \mathbf{X} > \delta \qquad \forall \mathbf{X} \in \mathbf{F}$$

Define

$$G(\mathbf{W}) = \frac{\mathbf{W}^* \cdot \mathbf{W}}{|\mathbf{W}|}$$

and note that $G(\mathbf{W})$ is the cosine of the angle between $\mathbf{W}$ and $\mathbf{W}^*$. Since $|\mathbf{W}^*| = 1$, we can say that

$$G(\mathbf{W}) \leq 1. \qquad (3.1)$$

Consider the behaviour of $G(\mathbf{W})$ through ADD .
Firstly, we can see how the numerator behaves:

$$
\begin{aligned}
\mathbf{W}^* \cdot \mathbf{W}_{t+1} &= \mathbf{W}^* \cdot (\mathbf{W}_t + \mathbf{X}) \\
&= \mathbf{W}^* \cdot \mathbf{W}_t + \mathbf{W}^* \cdot \mathbf{X} \\
&\geq \mathbf{W}^* \cdot \mathbf{W}_t + \delta
\end{aligned}
$$

since $\mathbf{W}^* \cdot \mathbf{X} > \delta$.
Hence, after the nth. application of ADD we have

$$\mathbf{W}^* \cdot \mathbf{W}_n \geq \delta \qquad (3.2)$$

Now we can consider the denominator, and since $\mathbf{W}^* \cdot \mathbf{X}$ must be negative, else the program would not go through ADD , we can say

$$
\begin{aligned}
|\mathbf{W}_{t+1}|^2 &= \mathbf{W}_{t+1} \cdot \mathbf{W}_{t+1} \\
&= (\mathbf{W}_t + \mathbf{X}) \cdot (\mathbf{W}_t + \mathbf{X}) \\
&= |\mathbf{W}_t|^2 + 2\mathbf{W}_t \cdot \mathbf{X} + |\mathbf{X}|^2
\end{aligned}
$$

However, we know that $\mathbf{W} \cdot \mathbf{X}$ must be negative, otherwise we would not be going through ADD , and we also know that $|\mathbf{X}|$, so we can write

$$< |\mathbf{W}_t|^2 + 1$$

(notice the = has become a <) and after the nth application of ADD ,

$$|\mathbf{W}_n|^2 < n \qquad (3.3)$$

Combining equations (3.2) and (3.3) gives us

$$\mathbf{G}(\mathbf{W}_n) \;=\; \frac{\mathbf{W}^* \cdot \mathbf{W}_n}{|\mathbf{W}_n|}$$

$$> \frac{n\delta}{\sqrt{n}}$$

but we already know that $\mathbf{G}(\mathbf{W}) \le 1$, so we can write

$$\sqrt{n}\delta \le 1$$

i.e.

$$n \le 1/\delta^2 \qquad (3.4)$$

Equation (3.4) is our proof: let us consider what it says. In the perceptron algorithm, we only go to TEST if $\mathbf{W} \cdot \mathbf{X} > 0$. We have chosen a small fixed number δ, such that $\delta > 0$ and $\mathbf{W} \cdot \mathbf{X} > \delta$. Equation (3.4) then says that we can make δ as small as we like, but the number of times, n, that we go to ADD will still be *finite*, and will be $\le 1/\delta^2$. In other words, eventually the perceptron will learn a weight vector $\mathbf{W}$ that partitions the space successfully, so that patterns from $\mathbf{F}^+$ are responded to with a positive output and patterns from $\mathbf{F}^-$ produce a negative output.

3.6   LIMITATIONS OF PERCEPTRONS

There are limitations to the capabilities of perceptrons, however. We have said before that they will learn the solution, *if there is a solution to be found*. To examine this in more detail, notice that the perceptron is trying to find the straight line that separates classes. It can separate classes that lie on either side of a straight line easily enough, but there are many situations where the division between

Table 3.1 The exclusive-or function table.

X	Y	Z
0	0	0
0	1	1
1	0	1
1	1	0

classes is much more complex. Consider the case of the exclusive-or (XOR) problem. The XOR logic function has two inputs and one output, shown in figure 3.9. It produces an output only if either one or the other of the inputs is on, but not if both are off or both are on. Representing on by 1, and off by 0, we can write this in a table as shown in table 3.1.

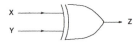

Figure 3.9 The exclusive-or logic symbol.

We can consider this as a problem that we want the perceptron to learn to solve: output a 1 if the X is on and Y is off, or is Y is on and X is off, otherwise output a 0. It appears to be a simple enough problem.

We can draw it in pattern space as shown in figure 3.10. The x-axis represents the value of X, the y-axis the value of Y. The heavily shaded circles represent the inputs that produce an output of 1, whilst the lighter circles show the inputs that produce an output of 0. Considering the heavily shaded circles and lightly shaded circles as separate classes, we find we *cannot* draw a straight line to separate the two classes (find a ruler, and try it!). Such patterns, as we have seen before in Chapter 2, are known as *linearly inseparable* since no straight line can divide them up successfully. Since we cannot divide

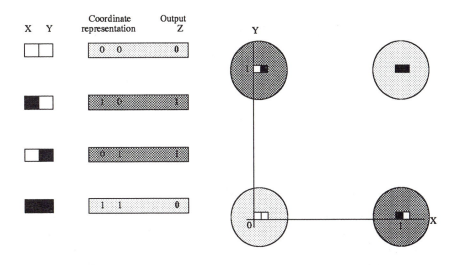

Figure 3.10 The XOR problem in pattern space.

them with a single straight line, the perceptron will not be able to find any such line either, and so cannot solve such a problem. In fact, a single-layer perceptron cannot solve any problem that is linearly inseparable.

3.7 THE END OF THE LINE?

The failure of the perceptron to successfully solve apparently simple problems such as the XOR one was first demonstrated by Minsky and Papert in their influential book *Perceptrons*. This book contained a detailed analysis of the capabilities and limitations of perceptrons; however, the demonstration that perceptrons could only do linearly separable problems was regarded as a mortal blow to the area, and the majority of the scientific community resolutely walked away.

3.7.1   Conclusions

The single-layer perceptron has shown great success for such a simple model. It has exhibited the features of learning that we wanted to realise in a system, and has shown that it is able to distinguish between different classes of objects if they are linearly separable in pattern space. What we need is a way to overcome the restraint of linear separability, whilst still retaining the basic features of the model and its overall simplicity. The improvement necessary first caught large-scale scientific attention in 1986 when Rumelhart and McClelland proposed their improved version, called the *multilayer* perceptron.

3.7.2   A Pause in History

One of the advantages in a book is that time is an illusion—one page turn can take you forward twenty years. In a neural network book, this is an advantage. Not much happened in the area after Minsky and Papert published their book in 1969, until Rumelhart and McClelland produced an improvement in 1986 which fused the perceptron idea with some modern adaptations and caused an explosion of interest in the field. If the McCulloch-Pitts neuron was the father of modern neural computing, then Rumelhart's multilayer perceptron is its child prodigy.

 Summary

- Perceptron—artificial neuron.
- Takes weighted sum of inputs, outputs +1 if greater than threshold else outputs 0.

- Hebbian learning (increasing effectiveness of active junctions) is predominant approach.
- Learning corresponds to adjusting the values of the weights.
- Feedforward supervised networks.
- Can use $+1, -1$ instead of $0, 1$ values.
- Can only solve problems that are linearly separable—therefore fails on XOR.

FURTHER READING

1. *Parallel Distributed Processing*, Volume 1. J. L. McClelland & D. E. Rumelhart. MIT Bradford Press, 1986. An excellent, broad-ranging book that covers many areas of neural networks. It was the book that signalled the resurgence of interest in neural systems.

2. *Organization of Behaviour*. Donald Hebb. 1949. Contains Hebb's original ideas regarding learning by reinforcement of active neurons.

3. *Perceptrons*. M. Minsky & S. Papert. MIT Press 1969. The criticisms of single-layer perceptrons are laid out in this book. A very interesting read, if a little too mathematical in places for some tastes.

4

The Multilayer Perceptron

4.1   INTRODUCTION

This chapter explores aspects of the multilayer perceptron, describing the modifications that need to be made to the basic model neuron in order to be able to solve more complex problems. The derivation of the learning rule is given and explained in full, and examples and applications of the network demonstrate its capabilities and potential.

4.2   ALTERING THE PERCEPTRON MODEL

4.2.1   The Problem

How are we to overcome the problem of being unable to solve linearly inseparable problems with our perceptron? An initial approach would be to use more than one perceptron, each set up to identify small, linearly separable sections of the inputs, then combining their outputs into another perceptron, which would produce a final indication of the class to which the input belongs. This approach to the XOR problem is shown in figure 4.1.

This seems fine on first examination, but a moment's thought will show that this arrangement of perceptrons in layers will be unable to learn. Each neuron in the structure still takes the weighted sum of its inputs, thresholds it, and outputs either a one or a zero. For the perceptrons in the first layer, the inputs come from the actual inputs to the network, while the perceptrons in the second layer take as their inputs the outputs from the first layer. This means

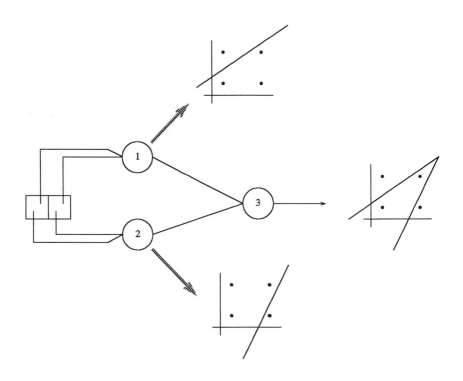

Figure 4.1 Combining perceptrons can solve the XOR problem: perceptron 1 detects when the pattern corresponding to (0,1) is present, and the other detects when (1,0) is there. Combined, these two facts allow perceptron 3 to classify the input correctly. They have to be set up correctly in the first place, however; they cannot learn to produce this classification.

that the perceptrons in the second layer do not know which of the real inputs were on or not; they are only aware of input from the first layer. Since learning corresponds to strengthening the connections between active inputs and active units (refer to section 3.3), it is impossible to strengthen the correct parts of the network, since the actual inputs are effectively masked off from the output units by the intermediate layer. The two-state neuron, being "on" or "off", gives us no indication of the scale by which we need to adjust the weights, and so we cannot make a reasonable adjustment. Weighted inputs that only just turn a neuron on should not be altered to the same extent as those in which the neuron is definitely turned on, but we have no way of finding out what the situation is. In other words, the hard-limiting threshold function (figure 3.3) removes the information that is needed if the network is to successfully learn. This difficulty is known as the *credit assignment* problem, since it means that the network is unable to determine which of the input weights should be increased and which should not, and so is unable to work out what changes should be made to produce a better solution next time.

4.2.2   The Solution

The way around the difficulty imposed by using the step function as the thresholding process is to adjust it slightly, and use a slightly different non-linearity. If we smooth it out, so that it more or less turns on or off, as before, but has a sloping region in the middle that will give us some information on the inputs, we will be able to determine when we need to strengthen or weaken the relevant weights. This means that the network will be able to learn, as required. A couple of possibilities for the new thresholding function are shown in figure 4.2.

In both cases, the value of the output will be practically one if the weighted sum exceeds the threshold by a lot, and conversely, it will be practically zero if the weighted sum is much less than the threshold value. However, in the case when the threshold and the weighted sum are almost the same, the output from the neuron will have a value somewhere between the two extremes. This means that

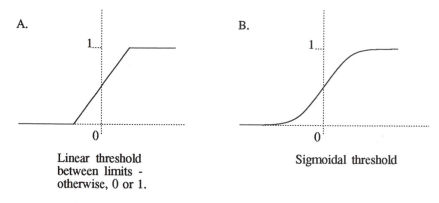

A.   B.

Linear threshold   Sigmoidal threshold
between limits -
otherwise, 0 or 1.

Figure 4.2 Two possible thresholding functions.

the output from the neuron is able to be related to its inputs in a more useful and informative way.

Notice that we have altered our model to try and overcome a particular difficulty by tracing the root of the problem, the hard-limiting thresholding that masks the inputs from the outputs, and then adjusting the model so that this can be solved. We have kept many of the essential features the same; each neuron still calculates the weighted sum, and thresholds it. However the input is now not simply on or off, but lies within a range, although the thresholding function that we are using approximates to the step function in many ways, especially at the extremes of its range. The solution that we have adopted is therefore one tailored to our particular problem, and it would be foolish of us to say that *real* biological neurons also work in this way. We are looking at an interesting construction of model neurons, and *not* at a small version of a real brain. This may appear obvious to the reader, but it is surprising how many false claims are made about models that have their roots in biological systems, and a timely reminder can do no harm.

We have to use a non-linear thresholding function, since layers of perceptron units using linear functions are no more powerful than a suitably chosen single layer. This is because each layer would be performing a purely linear operation on its inputs, which could be condensed into one operation. This is easiest to see with a simple example. Changing scale is a linear operation, since all things are affected by an equal amount. If a network scaled the input by 5 times in the first layer, and by 2 times in the second, that is exactly equivalent to one layer scaling the whole thing by 10 times.

4.3   THE NEW MODEL

The adapted perceptron units are arranged in layers, and so the new model is naturally enough termed the *multilayer perceptron*. The basic details are shown in figure 4.3.

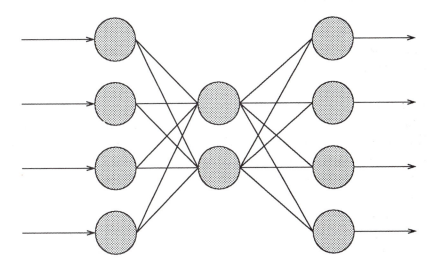

Figure 4.3 The multilayer perceptron: our new model.

Our new model has three layers; an input layer, an output layer, and a layer in between, not connected directly to the input or the output, and so called the *hidden* layer. Each unit in the hidden layer and the output layer is like a perceptron unit, except that the thresholding function is the one shown in figure 4.2, the sigmoid function B and not the step function as before. The units in the input layer serve to distribute the values they receive to the next layer, and so do not perform a weighted sum or threshold. Because we have modified the single-layer perceptron by changing the non-linearity from a step function to a sigmoid function, and added a hidden layer, we are forced to alter our learning rule as well. We now have a network that should be able to learn to recognise more complex things; let us examine the new learning rule in more detail.

4.4   THE NEW LEARNING RULE

The learning rule for multilayer perceptrons is called the "gener-alised delta rule", or the "backpropagation rule", and was suggested in 1986 by Rumelhart, McClelland, and Williams. It signalled the renaissance of the whole subject. It was later found that Parker had published similar results in 1982, and then Werbos was shown to have done the work in 1974. Such is the nature of science, however; groups working in diverse fields cannot keep up with all the advances in other areas, and so there is often duplication of effort. However, Rumelhart and McClelland are credited with reviving the perceptron since they not only developed the rule independently to the earlier claims, but used it to produce multilayer networks that they investi-gated and characterised. Their book, *Parallel Distributed Processing* is still one of the most important books in the field.

The operation of the network is similar to that of the single-layer perceptron, in that we show the net a pattern and calculate its re-sponse. Comparison with the desired response enables the weights to be altered so that the network can produce a more accurate output next time. The learning rule provides the method for adjusting the weights in the network, and, as we saw earlier in the chapter, the simple rule used in the single-layer perceptron will not work for mul-tilayer networks. However, the use of the sigmoid function means that enough information about the output is available to units in earlier layers, so that these units can have their weights adjusted so as to decrease the error next time.

The learning rule is a little more complex than the previous one, however, and we can best understand it by considering how the net behaves as patterns are taught to it. When we show the untrained network an input pattern, it will produce any random output. We need to define an error function that represents the difference between the network's current output and the correct output that we want it to produce. Because we need to know the "correct" pattern, this type of learning is known as "supervised learning". In order to learn successfully we want to make the output of the net approach the desired output, that is, we want to continually reduce the value of this error function. This is achieved by adjusting the weights on the links between the units, and the generalised delta rule does this by calculating the value of the error function for that particular input, and then back-propagating (hence the name!) the error from one layer to the previous one. Each unit in the net has its weights adjusted so that it reduces the value of the error function; for units actually on the output, their output and the desired output is known, so adjusting the weights is relatively simple, but for units in the middle layer, the adjustment is not so obvious. Intuitively, we might guess that the hidden units that are connected to outputs with a large error should have their weights adjusted a lot, while units that feed almost correct outputs should not be altered much. In fact, the mathematics shows that the weights for a particular node should be adjusted in direct proportion to the error in the units to which it is connected: that is why back-propagating these errors through the net allows the weights between all the layers to be correctly adjusted. In this way the error function is reduced and the network learns.

4.4.1   The Mathematics

Firstly, the notation used is as follows:

E_p is the error function for pattern p, t_{pj} represents the target output for pattern p on node j, whilst o_{pj} represents the actual output at that node. w_{ij} is the weight from node i to node j.

Let us define the error function to be proportional to the square of the difference between the actual and desired output, for all the patterns to be learnt.

$$E_p = \frac{1}{2}\sum_j (t_{pj} - o_{pj})^2 \tag{4.1}$$

The $\frac{1}{2}$ makes the maths a bit simpler, and brings this specific error function into line with other similar measures.

The activation of each unit j, for pattern p, can be written as

$$net_{pj} = \sum_i w_{ij} o_{pi} \tag{4.2}$$

i.e. simply the weighted sum, as in the single-layer perceptron.

The output from each unit j is the threshold function f_j acting on the weighted sum. In the perceptron, this was the step function; in the multilayer perceptron, it is usually the sigmoid function, although any continuously differentiable monotonic function can be used.

$$o_{pj} = f_j(net_{pj}) \tag{4.3}$$

We can write

$$\frac{\partial E_p}{\partial w_{ij}} = \frac{\partial E_p}{\partial net_{pj}} \frac{\partial net_{pj}}{\partial w_{ij}} \tag{4.4}$$

by the chain rule.

Looking at the second term in (4.4), and substituting in (4.2),

$$\frac{\partial net_{pj}}{\partial w_{ij}} = \frac{\partial}{\partial w_{ij}} \sum_k w_{kj} o_{pk}$$

$$= \sum_k \frac{\partial w_{jk}}{\partial w_{ij}} o_{pk}$$

$$= o_{pi} \tag{4.5}$$

since $\dfrac{\partial w_{kj}}{\partial w_{ij}} = 0$ except when $k = i$ when it equals 1.

We can define the change in error as a function of the change in the net inputs to a unit as

$$-\frac{\partial E_p}{\partial net_{pj}} = \delta_{pj} \tag{4.6}$$

and so (4.4) becomes

$$-\frac{\partial E_p}{\partial w_{ij}} = \delta_{pj} o_{pi} \tag{4.7}$$

Decreasing the value of E_p therefore means making the weight changes proportional to $\delta_{pj} o_{pi}$, i.e.

$$\Delta_p w_{ij} = \eta \delta_{pj} o_{pi} \tag{4.8}$$

We now need to know what δ_{pj} is for each of the units—if we know this, then we can decrease E. Using (4.6) and the chain rule, we can write

$$\delta_{pj} = -\frac{\partial E_p}{\partial net_{pj}} = -\frac{\partial E_p}{\partial o_{pj}} \frac{\partial o_{pj}}{\partial net_{pj}} \tag{4.9}$$

Consider the second term, and from (4.3),

$$\frac{\partial o_{pj}}{\partial net_{pj}} = f'_j(net_{pj}) \tag{4.10}$$

Consider now the first term in (4.9). From (4.1), we can differentiate E_p with respect to o_{pj}, giving

$$\frac{\partial E_p}{\partial o_{pj}} = -(t_{pj} - o_{pj}) \tag{4.11}$$

Thus

$$\delta_{pj} = f'_j(net_{pj})(t_{pj} - o_{pj}) \tag{4.12}$$

This is useful for the output units, since the target and output are both available, but not for the hidden units, since their targets are not known.

So, if unit j is not an output unit, we can write, by the chain rule again, that

$$\begin{aligned}
\frac{\partial E_p}{\partial o_{pj}} &= \sum_k \frac{\partial E_p}{\partial net_{pk}} \frac{\partial net_{pk}}{\partial o_{pj}} \\
&= \sum_k \frac{\partial E_p}{\partial net_{pk}} \frac{\partial}{\partial o_{pj}} \sum_i w_{ik} o_{pi} \tag{4.13} \\
&= -\sum_k \delta_{pk} w_{jk} \tag{4.14}
\end{aligned}$$

using (4.2) and (4.6), and noticing that the sum drops out since the partial differential is non-zero for only one value, just as in (4.5). Substituting (4.14) in (4.9), we get finally

$$\delta_{pj} = f'_j(net_{pj}) \sum_k \delta_{pk} w_{jk} \qquad (4.15)$$

This equation represents the change in the error function, with respect to the weights in the network. This provides a method for changing the error function so as to be sure of reducing it. The function is proportional to the errors δ_{pk} in subsequent units, so the error has to be calculated in the output units first (given by (4.12)) and then passed back through the net to the earlier units to allow them to alter their connection weights. It is the passing back of this error value that leads to the networks being referred to as *back-propagation* networks. Equations (4.12) and (4.15) together define how we can train our multilayer networks.

One advantage of using the sigmoid function as the non-linear threshold function is that it is quite like the step function, and so should demonstrate behaviour of a similar nature. The sigmoid function is defined as

$$f(net) = 1/(1 + e^{-k\ net})$$

and has the range $0 < f(net) < 1$. k is a positive constant that controls the "spread" of the function—large values of k squash the function until as $k \to \infty, f(net) \to$ Heaviside function. It also acts as an automatic gain control, since for small input signals the slope is quite steep and so the function is changing quite rapidly, producing a large gain. For large inputs, the slope and thus the gain is much less. This means that the network can accept large inputs and still remain sensitive to small changes.

A major reason for its use is that it has a simple derivative, however, and this makes the implementation of the back-propagation system much easier. Given that the output of a unit, o_{pj} is given by

$$o_{pj} = f(net) = 1/(1 + e^{-k\ net})$$

the derivative with respect to that unit, $f'(net)$, is given by

$$f'(net) \quad = \quad ke^{-k\ net}/(1 + e^{-k\ net})^2$$

$$= \ k \ f(net)(1 - f(net))$$
$$= \ k \ o_{pj}(1 - o_{pj})$$

The derivative is therefore a simple function of the outputs.

4.5   THE MULTILAYER PERCEPTRON ALGORITHM

The algorithm for the multilayer perceptron that implements the back-propagation training rule is shown below. It requires the units to have thresholding non-linear functions that are continuously differentiable, i.e. smooth everywhere. We have assumed the use of the sigmoid function, $f(net) = 1/(1 + e^{-k \ net})$ since it has a simple derivative.

Multilayer Perceptron Learning Algorithm

---

1. Initialise weights and thresholds
Set all weights and thresholds to small random values.
2. Present input and desired output
Present input $X_p = x_0, x_1, x_2, \ldots, x_{n-1}$ and target output $T_p = t_0, t_1, \ldots, t_{m-1}$ where n is the number of input nodes and m is the number of output nodes. Set w_0 to be $-\theta$, the bias, and x_0 to be always 1. For pattern association, X_p and T_p represent the patterns to be associated. For classification, T_p is set to zero except for one element set to 1 that corresponds to the class that X_p is in.
3. Calculate actual output
Each layer calculates

$$y_{pj} = f \left[\sum_{i=0}^{n-1} w_i x_i \right]$$

and passes that as input to the next layer. The final layer outputs values o_{pj}.
4. Adapt weights

Start from the output layer, and work backwards.

$$w_{ij}(t+1) = w_{ij}(t) + \eta \delta_{pj} o_{pj}$$

$w_{ij}(t)$ represents the weights from node i to node j at time t, η is a gain term, and δ_{pj} is an error term for pattern p on node j.

For output units

$$\delta_{pj} = k o_{pj}(1 - o_{pj})(t_{pj} - o_{pj})$$

For hidden units

$$\delta_{pj} = k o_{pj}(1 - o_{pj})\sum_k \delta_{pk} w_{jk}$$

where the sum is over the k nodes in the layer above node j.

4.6   THE XOR PROBLEM REVISITED

In the previous chapter, we saw how the single-layer perceptron was unable to solve the exclusive-or problem. Since this problem showed the limitations of single-layer perceptrons, it has become the yardstick by which the performance of many new neural systems are judged, and many features of the behaviour of multilayer perceptrons are revealed by it.

To quickly review it, the problem is to classify the following correctly:

Input	Output
00	0
01	1
10	1
11	0

The first test of the multilayer perceptron is to see if we can produce a network that can solve this problem; the two-layer net shown in figure 4.4 is able to produce the correct output. It has a three-layer structure, with two input units (as we might expect since there are two variables in the problem), one unit in the hidden layer, and one output unit. The connection weights are shown on the links, and the threshold of each unit is shown inside the unit. As far as the output unit is concerned, the hidden unit is no different from either of the input units, and simply provides another input.

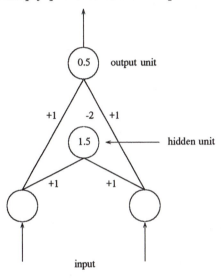

Figure 4.4 A solution to the XOR problem.

Notice that the hidden unit's threshold of 1.5 means that it is off unless turned on by both inputs being on. This is an important point to note. It is interesting to look at the behaviour of the network as it solves the XOR problem. When both inputs are off (00), the hidden unit is also off, and there is no net input to the output unit, which therefore remains off. When the right input only is on (01), the hidden unit does not receive enough net input to turn it on, so it remains off. The output unit sees a net input of +1, which exceeds its threshold, and so turns it on. The same happens when the left unit only (10) is on. When both input units are on (11) the hidden unit receives a net input of +2, which exceeds its threshold value, and so it turns on. The output unit now sees a net input of +1 from each of the input units, making +2, and −2 from the hidden unit, making 0 in all. This is less than the threshold, and so the unit is off. This can be summarised in the table below.

Input	Hidden Unit	Output
00	0	0
01	0	1
10	0	1
11	1	0

Considering the hidden unit, we can see that it is detecting when *both* the inputs are on, since this is the only condition under which it turns on. Since each of the input units detect when their inputs are on, the output unit is fed with three items of information: if the left input is on, if the right input is on, and if both the left and the right inputs are on. Since the output unit treats the hidden unit as another input unit, the apparent input patterns it receives are now dissimilar enough for the classification to be learnt.

The hidden unit acts as a *feature detector*, detecting when both the inputs are on. It can be viewed as recoding the basic inputs so that the network can learn the required mapping of input patterns to output ones. This recoding, or *internal representation*, is critical to the functioning of the network. Given enough hidden units, it is possible to form internal representations of any input pattern such that the output units are able to produce the correct response for a specific input.

The generalised delta rule provides a method for teaching multilayer perceptron networks, producing the necessary internal representations on the hidden nodes. It is unlikely that the weights produced by a taught network would be as simple as those shown above, but the same principles hold. Figure 4.5 shows another solution to the XOR problem.

Multilayer perceptrons can appear in all shapes and sizes, with the same learning rule for them all. This means that it is possible to

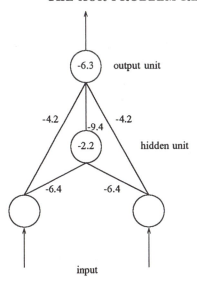

Figure 4.5 Weights and thresholds of a network that has learnt to solve the XOR problem.

produce different network topologies to solve the same problem. One of the more interesting cases is when there is no direct connection from the input to the output. This and the corresponding XOR solution are shown in figure 4.6. The right-hand hidden unit detects when both inputs are on, and ensures that the output unit gets a net input of zero. When only one of the inputs is on, the left-hand hidden unit is on, turning on the output unit. When both inputs are off, the hidden units are inactive and so the output unit is off.

The learning rule is not guaranteed to produce convergence, however, and it is possible for the network to fall into a situation in which it is unable to learn the correct output.

The network shown in figure 4.7 will correctly respond to the input patterns 00 and 10, but fails to produce the correct output for the patterns 01 or 11. The right-hand input turns on both hidden units. These produce a net input of 0.8 at the output unit, exactly the same as the threshold value. Since the thresholding function is the sigmoid, this gives an output value of 0.5. This situation is stable

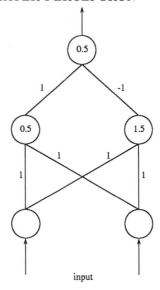

Figure 4.6 An XOR-solving network with no direct input-output connections.

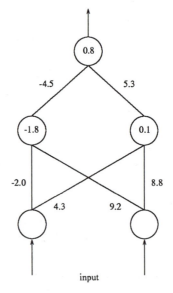

Figure 4.7 A stable solution that does not work.

and does not alter with further training. This *local minimum* occurs infrequently—about 1% of the time in the XOR problem.

Another minor problem can occur in training networks with the generalised delta rule. Since the weight changes are proportional to the weights themselves, if the system starts off with equal weights then non-equal weights can never be developed, and so the net cannot settle into the non-symmetric solution that may be required.

4.7   VISUALISING NETWORK BEHAVIOUR

Having looked at the generalised delta rule informally, and mathematically, and having examined the multilayer perceptron solving the XOR problem, we need a method of visualising what is going on in the network. The mathematical analysis of the networks does provide a convenient and useful approach to the visualisation of their behaviour. As we have seen, the network computes an error or energy function, $E_p = -\frac{1}{2}\sum(t_{pj} - o_{pj})^2$ which represents the amount by which the output of the net differs from the required output. Large differences correspond to large energies, whilst small differences correspond to small energies. Since the output of the net is related to the weights between the units and the input applied, the energy is therefore a function of the weights and inputs to the network. We can draw a graph of the energy function showing how varying the weights affects the energy, for a fixed input pattern. Considering this for a moment, this means that if we imagine a very odd network in which we can only vary one weight, we can plot a graph of the energy function for a particular pattern versus the weight, which may look something like figure 4.8.

If we extend our thinking so that we can vary two weights, we will then have two axes for the weights, and the graph of the energy function would appear, for example, like figure 4.9.

We obtain a three-dimensional graph with two weight axes and one energy axis; if we allowed another weight to vary, then we would have *another* axis to add, which would be difficult! In general, we can adjust all the weights in a network, and there may be very many of them, giving a multidimensional energy function, which we cannot

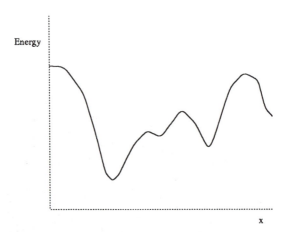

Figure 4.8 The energy function in one dimension, as we vary one weight, for a fixed pattern.

draw. However, it is useful to consider it as best we can, as a surface in 3-d, and just keep reminding ourselves that it is in fact multidimensional. Our understanding of the higher dimensioned case is helped greatly by the analogies that we can visualise easily in the 3-d situation. This energy surface is a rippling landscape of hills and valleys, wells and mountains, with points of minimum energy corresponding to the wells and maximum energy found on the peaks. The generalised delta rule aims to minimise the error function E by adjusting the weights in the network so that they correspond to those at which the energy surface is lowest. It does this by a method known as *gradient descent*, where the energy function is calculated, and changes are made in the steepest downward direction. This is guaranteed to find a solution in cases where the energy landscape is simple. Each possible solution is represented as a hollow, or a basin, in the landscape. These *basins of attraction*, as they are known, represent the solutions to the values of the weights that produce the correct output from a given input. Remember that these basins are

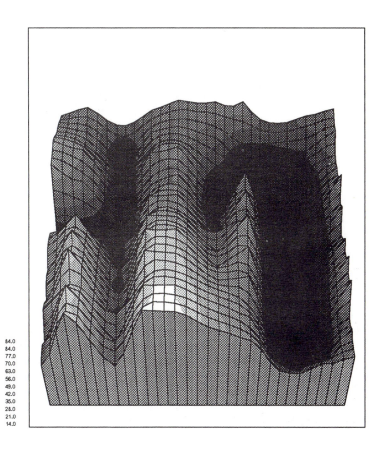

	ABOVE	84.0
	77.0 -	84.0
	70.0 -	77.0
	63.0 -	70.0
	56.0 -	63.0
	49.0 -	56.0
	42.0 -	49.0
	35.0 -	42.0
	28.0 -	35.0
	21.0 -	28.0
	14.0 -	21.0
	BELOW	14.0

Figure 4.9 The energy function in two dimensions. Notice the ravine on the right: starting in the middle near the front and going downhill may take you either straight down to the ravine floor, or around the sharp peak back right, depending on how often you work out which way is down, and where you start from. Notice also that the valley on the left has lots of small hollows in its floor. These local minima can trap the solution and prevent it reaching the deeper point which occurs about halfway along.

actually many-dimensional, but we can only draw them in 3-d.

It is easiest to visualise this energy surface as a large, stretchy, rubber sheet that is initially flat. The basins of attraction are formed by placing heavy balls on the sheet; the sheet deforms downwards creating a well. The bottom of the well represents the low energy solution that the network has learnt.

We can also imagine a many-dimensioned space in which each axis represented one particular weight—in this case one point in the space would represent one unique combination of possible weight values that the network could have. This space is known, sensibly enough, as the *weight space*. Our example of energy space described how we could visualise the energy changing as we varied the weights for a particular pattern—however, we could have just as easily imagined how the energy would change as we varied the input patterns for a particular fixed set of weights. Each point in the weight space therefore defines a different energy landscape, where the variables are the patterns and their corresponding energies. This behaviour is shown in figure 4.10.

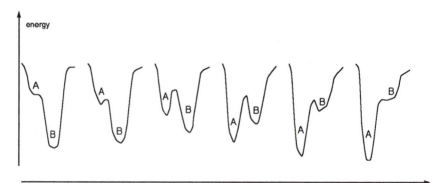

adjusting weight in direction that favours the storage of pattern A

Figure 4.10 Diagram showing how changing the weights in a network alters the energy landscape. In this case, the weight change from left to right favours pattern A since it lowers the energy of that pattern at the expense of pattern B.

Many of the features associated with multilayer perceptrons are easiest to understand if they are considered in terms of the energy landscape.

4.8   MULTILAYER PERCEPTRONS AS CLASSIFIERS

We have already considered how the multilayer perceptron copes with the complicated, linearly inseparable XOR problem; now we consider the more general case. The single-layer perceptron is limited to calculating a single plane of separation between classes, which is why it fails on problems such as the XOR which are more complicated. We discussed earlier how a two-layer device could, in principle, solve the XOR problem. Consider a net of three perceptron devices as shown in figure 4.11.

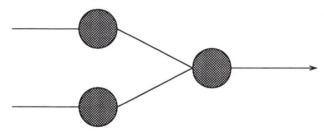

Figure 4.11 Two perceptron units can be combined to produce input for a third.

If the unit in the second layer has its threshold set so that it turns on only when both of the first-layer units are on, it is performing a logical AND operation. Since each of the units in the first layer defines a line in pattern space, the second unit produces a classification based on a combination of these lines. If one unit is set to respond with a 1 if the input is above its decision line, and the other responds with a 1 if the same input is below its decision line, then the second layer produces a solution as shown in figure 4.12, producing a 1 if it is above line 1 *and* below line 2.

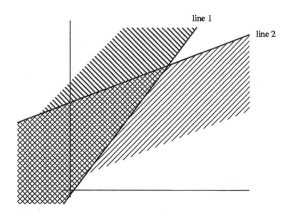

Figure 4.12 3 perceptrons: the decision region produced by combining 2 perceptrons with another.

More than two units can be used in the first layer, which produces pattern space partitioning that is a combination of more than 2 lines. All regions produced in this way are known as *convex regions* or *convex hulls*. A convex hull is a region in which any point can be connected to any other by a straight line that does not cross the boundary of the region. Regions can be closed or open; a closed region has a boundary all around it, as in shapes such as a triangle or a circle, whilst an open region does not, as between two parallel lines. Examples of closed and open convex regions are shown in figure 4.13.

The addition of more perceptron units in the first layer allows us to define more and more edges—from the points we have made above, it is obvious that the total number of sides that we can have in our regions will be at most equal to the number of units in the first layer, and that the regions defined will still be convex.

However, if we add another layer of perceptrons, the units in this layer will receive as inputs, not lines, but convex hulls, and the combinations of these are not necessarily convex, as shown in figure 4.14. The combinations of these convex regions may intersect, overlap, or be separate from each other, producing arbitrary shapes.

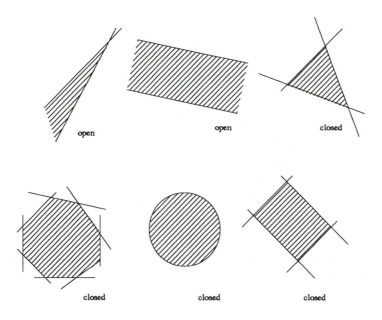

Figure 4.13 Examples of closed and open convex hulls.

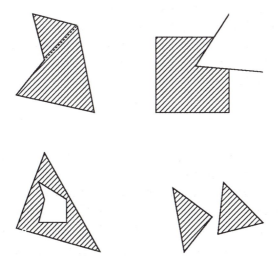

Figure 4.14 Examples of arbitrary regions formed by the combination of various convex regions.

Three layers of perceptron units can therefore form arbitrarily complex shapes, and are capable of separating any classes. The complexity of the shapes is limited by the number of nodes in the network, since these define the number of edges that we can have. The arbitrary complexity of shapes that we can create, means that we never need more than three layers in a network, a statement that is referred to as the *Kolmogorov theorem*. This can be proved, with a bit of complex maths, but it will suffice to state it here. A summary of the perceptron's classification abilities is shown in figure 4.15.

The neural network literature is inconsistent when describing networks, since some authors refer to the number of layers of variable weights, whilst others describe the number of layers of nodes. This causes confusion since the nodes in the first layer, the input layer, merely distribute the inputs to subsequent layers, and do not perform any summation or thresholding themselves. To confuse matters further, some authors miss out these input nodes altogether when drawing diagrams! To try to clarify the situation: a multilayer network receives a number of inputs. These are distributed by a layer of input nodes that do not perform any summation or thresholding— these input nodes have only one input each, so it is clear which they are, and obviously pointless for them to sum their only input. These inputs are then passed along the first layer of adaptive weights to a layer of perceptron-like units, which do sum and threshold their inputs. This layer is able to produce classification lines in pattern space. The output from this layer is then passed to another layer of perceptron-like units via adaptable weights, and it is the output of this layer that forms convex hulls in pattern space. A further layer of perceptron-like units is reached by another set of adaptive weights, and the output of this layer is able to define any arbitrary shape in pattern space. Counting the number of active weight layers, or the number of active perceptron layers, this is a three-layer network. If the inactive set of input units is included, it can be called a four-layer network. The general trend is to use the former, since it is more descriptive. This is summarised in figure 4.16.

It has been known for a long time that layers of perceptrons would be able to do more than single ones, but until the generalised delta

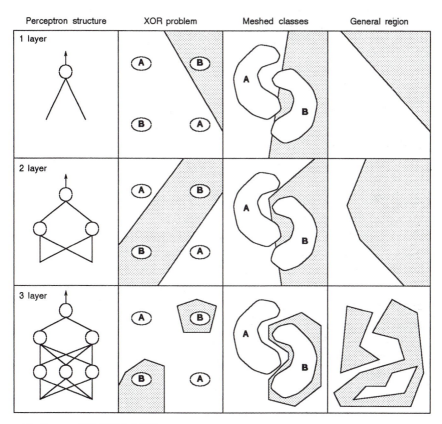

Perceptron structure	XOR problem	Meshed classes	General region
1 layer			
2 layer			
3 layer			

(after Lippmann, IEEE ASSP April 1987)

Figure 4.15 Neural networks and their corresponding decision regions.

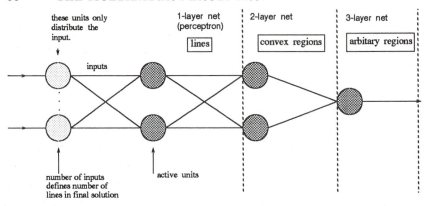

Figure 4.16 Summary of the boundaries formed by different numbers of perceptron layers.

rule was formulated there was no learning algorithm for such layered networks. The use of the sigmoidal non-linearity in the multilayer perceptron units transforms the straight line decision surface of the perceptron into a smooth curve, and so the regions formed are now also bounded by smooth curves, but the overall complexity of the shapes of the regions for two- and three-layer networks remains the same.

We can consider classifying patterns in another way. Any given input pattern must belong to one of the classes that we are considering, and so there is a mapping from the input to the required class. This mapping can be viewed as a function that transforms the input pattern into the correct output class, and we can consider that a network has learnt to perform correctly, if it can carry out this mapping. In fact, *any* function, no matter how complex, can be represented by a multilayer perceptron of no more than three layers; the inputs are fed through an input layer, a middle hidden layer, and an output layer. As we have already mentioned, this is known as the Kolmogorov representation theorem; it is an important result in that it proves that whatever is done in four or more layers could also be done in three. It therefore limits the number of layers that are necessary to represent an arbitrary function, but unfortunately it gives no indication as to how many units the network requires, how they should be connected, or how the weights between them should be set.

4.9   GENERALISATION

One of the major features of neural networks is their ability to generalise, that is, to successfully classify patterns that have not been previously presented. Multilayer perceptrons generalise by detecting features of the input pattern that have been learnt to be significant, and so coded into the internal units. Thus an unknown pattern is classified with others that share the same distinguishing features. This means that learning by example is a feasible proposition, since only a representative set of patterns have to be taught to the network, and the generalisation properties will allow similar inputs to be classified as well. It also means that noisy inputs will be classified, by virtue of their similarity with the pure input. It is this generalisation ability that allows multilayer perceptrons to perform more successfully on real-world problems than other pattern recognition or expert system methods.

In general, neural networks are good at interpolation, but not so good at extrapolation. They are able to detect the patterns that exist in the inputs they are given, and allow for intermediate states that have not been seen. However, inputs that are extensions of the range of patterns are less well classified, since there is little with which to compare them. Put another way, given an unseen pattern that is an intermediate mixture of two previously taught patterns, the net will classify it as an example of the predominant pattern. If the pattern does not correspond to anything similar to what the net has seen before, then classification will be much poorer.

4.10   FAULT TOLERANCE

Multilayer perceptron networks are intrinsically fault-tolerant, since they are distributed parallel processing elements, with each node contributing to the final output response. If a node or its weights are lost or damaged, recall is impaired in quality, but the distributed nature of the information means that damage has to be extensive before a network's response degrades badly. The network therefore demonstrates graceful degradation in performance rather than catastrophic failure.

They are also tolerant to noise due to their intrinsic ability to generalise from taught examples to corrupted versions of the original patterns.

Damage to a network, whether it takes the form of the loss of a few nodes or the incorporation of noise into the training data, can often be recovered from by relearning, and in these cases the recovery of the network is often very quick. This can be understood by examining figure 4.17. Convergence to the original solution was

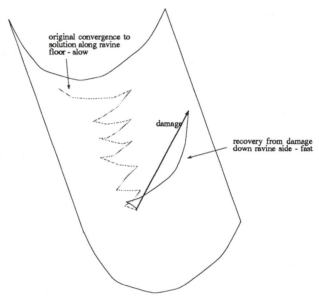

Figure 4.17 Diagram showing how recovery from damage can be achieved quickly.

along a valley floor, and so was slow. The damage done upsets the network, but it is quite likely to move it into a state that has a large gradient towards the correct solution, and so when relearning occurs, the net moves along this steep gradient and quickly recovers the original solution.

4.11   LEARNING DIFFICULTIES

The XOR problem demonstrates some of the difficulties associated with learning in multilayer perceptrons. Occasionally the network settles into a stable solution that does not provide the correct output. In these cases, the energy function is in a *local minimum*. This means that in every direction in which the network could move in the energy landscape, the energy is higher than at the current position. It may be that there is only a slight "lip" to cross before reaching an actual deeper minimum, but the network has no way of knowing this, since learning is accomplished by following the energy function down in the steepest direction, until it reaches the bottom of a well, at which point there is no direction to move in order to reduce the energy.

There are alternative approaches to minimising these occurrences, which are outlined below.

- Lowering the gain term

    If the rate at which the weights are altered is progressively decreased, then the gradient descent algorithm is able to achieve a better solution. If the gain term η is made large to begin with, large steps are taken across the weight and energy space towards the solution. As the gain is decreased, the network weights settle into a minimum energy configuration without overshooting the stable position, as the gradient descent takes smaller downhill steps. This approach enables the network to bypass local minima at first, then hopefully locate, and settle in, some deeper minima without oscillating wildly. However, the reduction in the gain term will mean that the network will take longer to converge.

- Addition of internal nodes

    Local minima can be considered to occur when two or more disjoint classes are categorised as the same. This amounts to a poor internal representation within the hidden units, and so adding more units to this layer will allow a better recoding of the inputs and lessen the occurrence of these minima.

- Momentum term

    The weight changes can be given some "momentum" by introducing an extra term into the weight adaptation equation that will produce a large change in the weight if the changes are currently large, and will decrease as the changes become less. This means that the network is less likely to get stuck in local minima early on, since the momentum term will push the changes over local increases in the energy function, following the overall downward trend. Momentum is of great assistance in speeding up convergence along shallow gradients, allowing the path the network takes towards the solution to pick up speed in the downhill direction. The energy landscape may consist of long gradually sloping ravines which finish at minima. Convergence along these ravines is slow, since the direction that has to be followed has only a slight gradient, and usually the algorithm oscillates across the ravine valley as it meanders towards a solution, as shown in figure 4.18. This is difficult to speed up without increasing the chance of overshooting the minima, but the addition of the momentum term is fairly successful.

    This momentum term can be written as follows:

    $$\delta_p w_{ji}(t+1) = w_{ji}(t) + \eta \delta_{pj} o_{pi} + \alpha \left(w_{ji}(t) - w_{ji}(t-1) \right)$$

    where α is the momentum factor, $0 < \alpha < 1$.

- Addition of noise

    If random noise is added, this perturbs the gradient descent algorithm from the line of steepest descent, and often this noise is enough to knock the system out of a local minimum. This approach has the advantage that

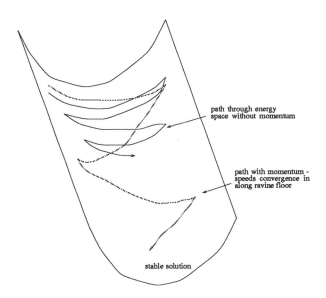

path through energy
space without momentum

path with momentum -
speeds convergence in
along ravine floor

stable solution

Figure 4.18 The addition of a momentum term can speed up convergence, especially along a ravine.

it takes very little extra computation time, and so is not noticeably slower than the direct gradient descent algorithm.

4.11.1 Other Learning Problems

One of the major criticisms of the multilayer perceptron is that it requires many presentations of the set of input patterns, and the repetition of the corresponding calculation and back-propagation of the errors for each pattern, before the network is able to settle into a stable solution. The method of gradient descent is intrinsically slow to converge in a complex landscape, due to the complexity of the energy surface. The addition of the momentum term, as discussed in the previous section, often speeds convergence, whilst another method is to alter the gain term η. Another alternative solution, which also helps to avoid spurious local minima, is to take account of second order effects in the gradient descent algorithm. However, the increased accuracy of the line of descent offered by this solution is offset by the additional computational complexity involved.

4.12   RADIAL BASIS FUNCTIONS

An enhancement to the standard multilayer perceptron techniques uses what are known as *radial basis functions*. These are a set of generally non-linear functions that are built up into one function that can partition the pattern space successfully. The usual multilayer perceptron builds its classifications from hyperplanes, defined by the weighted sums $\sum_j w_{ij}x_i$ which are arguments to non-linear functions, whereas the radial basis approach uses *hyperellipsoids* to partition the pattern space. These are defined by functions of the form $\phi(||x - y||)$ where $||\ldots||$ denotes some distance measure. We can intuitively see that this expression describes some sort of multi-dimensional ellipse, since it represents a function whose argument is related to a distance from a centre, y. The function s in k-dimensional space, which partitions the space, has elements s_k given by

$$s_k = \sum_{j=1}^{m} \lambda_{jk}\phi(||x - y_j||)$$

In other words, it is a *linear* combination of these basis functions.

The advantage of using the radial basis approach is that once the radial basis functions have been chosen, all that is left to determine are the coefficients λ_j for each, to allow them to partition the space correctly. Since these coefficients are added in a linear fashion, the problem is an exact one and has a guaranteed solution since there are no nasty local minima situations in which to fall. In effect, the radial basis functions have expanded the inputs into a higher-dimensional space where they are now linearly separable.

This approach is guaranteed to produce a function that fits all the data points, as long as there is a basis function for each input to be classified. Having one basis function for each input does mean that noisy or anomalous data points will also be classified, however, and these will tend to cause distortion. This noise distortion causes problems with generalisation; since the classification surface is not necessarily smooth, very similar inputs may find themselves assigned to very different classes. The solution to this is to reduce the number of basis functions to a level at which an acceptable fit to the data is still achieved. This means that the previously exact problem becomes one of linear optimisation, but this is not a complex technique, and the classification surface will be smooth between the data points.

The choice of which radial basis functions to use is usually made in one of two ways. In the absence of any knowledge about the data, the basis functions are chosen so that they fit points evenly distributed through the set of possible inputs. If we have some knowledge as to the overall structure of the inputs, then it is better to try and mirror that structure in the choice of functions. This is most easily achieved by choosing a subset of the input points, which should have a similar distribution to the overall input, as the points to be fitted.

The function ϕ is usually chosen to be a Gaussian function, i.e.

$$\phi(r) = e^{(-r^2)}$$

whilst the distance measure $|| \ldots ||$ is taken to be Euclidean:

$$||x - y|| = \sum_i (x_i - y_i)^2$$

where y represents the centre of the hyperellipse.

This can be represented in a network as shown in figure 4.19.

The y_{jk} terms in the first layer are fixed, and the input to the nodes on the hidden layer is given, in the case of the Euclidean distance measure, as

$$\sum_{i=1}^{n} (x_i - y_{jk})^2$$

This hidden layer is fully connected to the output layer by connections of strengths λ_{jk} and it is these that have to be linearly optimised.

The use of radial basis functions is becoming more popular, since they need only linear optimisation techniques, which provide a guaranteed, globally optimal solution. The difficulty in using them is in

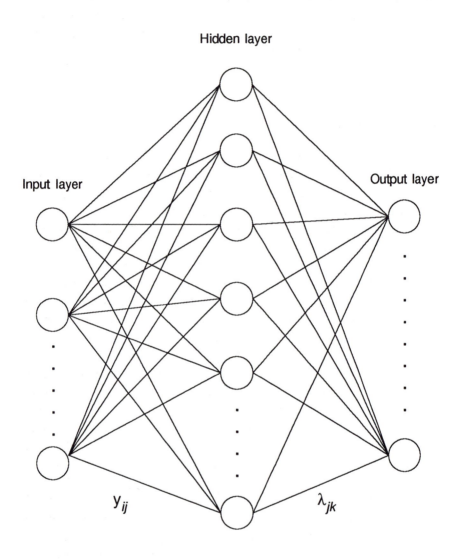

Figure 4.19 A feedforward network showing how it represents radial basis functions.

deciding on the set of basis functions to be used, in order to get an acceptable fit to the data. This is one of a number of techniques that essentially preprocess the data and transform it into a higher-dimensional space in which the classes are linearly separable.

4.13   APPLICATIONS

4.13.1   NETtalk

One of the most famous and influential network applications is called NETtalk, a multilayer perceptron that learns to pronounce English text, and was developed by Sejnowski and Rosenberg in 1987. It consists of 203 input units, 80 hidden units, and 26 output units, one for each phoneme—a phoneme is a basic sound in the language, from which all words are composed. This is shown in figure 4.20. A window seven letters wide is moved over the text, and the net learns to pronounce the middle letter. The windowing of the text before and after the pronounced character provides context sensitivity, since the sound of letters within a word is dependent on the word itself— for example, the "a" in "mean" is virtually silent, in "lamb" it is a short, sharp 'a', whilst in "class" it is an 'arr' sound.

The appealing feature of the network is that it appears to mimic the speech patterns of young children, producing an incoherent babble at first since the weights are random, then picking out the major features of the English language, namely the "ee-oo-ee-oo-ee" patterns that words make. (Listen to the overall sounds made when someone speaks. As they talk, their voice rises and falls in the same manner as the "ee-oo-ee-oo-ee" phrase—try saying it to get the full effect!) Repeated training produces more and more intelligible speech. The network achieves about 90% correct phoneme pronunciation, and its generalisation has also been investigated by training it on words from a dictionary, then testing on an unseen set. Again, about 90% correct pronunciation of phonemes was reported for the training set, with between 80% and 87% on the unseen set, increasing as the size of the training set was increased. The net was also resistant to damage in the form of random noise added to the weights, and showed a graceful degradation in performance.

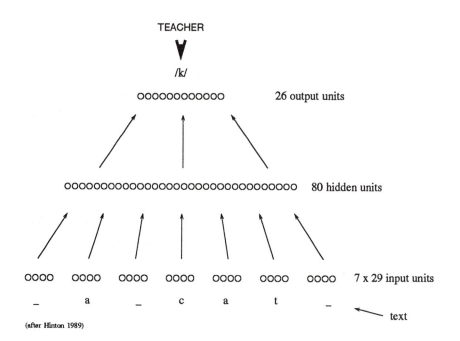

Figure 4.20 The layout of the NETtalk network.

4.13.2   Airline Marketing Tactician

Have you ever flown on holiday, and found the plane has had half a dozen empty seats? Worse for you, have you ever arrived at an airport to check in, only to find that your seat has been double-booked and you have to wait for the next flight? The difficulty this causes is one of the dilemmas facing airlines as they struggle to fill all the places on their planes, since empty seats are lost revenue. Knowing that a number of people will book and not show up, they overbook the seats, trusting that it will all work out in the end. The problem they have is that of accurately predicting demand for seats and the proportion of passengers that will not turn up (known as no-show passengers), so that they can set the overbooking limits to allow for these. This tactical marketing aims to maximise the profitability of the airline, who have many conflicting factors to take into account in their strategy. They want to sell as many seats at as high a price as possible, but realise that it is better to sell the seats cheaply than not at all. The loss of money from an empty seat means that they have to slightly overbook to compensate, but the cost of a denied boarding for an unlucky customer is much larger than the cost of flying with an empty seat.

The Airline Marketing Tactician (AMT) is a two-stage procedure that assists the airline; the first stage consists of a multilayer perceptron that produces forecasts of seat demand, and the second stage allocates airline resources to meet these projected demands using standard optimisation techniques. This two-stage system is preferable to the direct allocation of resources by the network in a single-stage process, since the forecasts that the network produces can be checked against the actual demand, and these provide the justification for the allocation decisions. The two-stage process is therefore much easier for humans to follow and analyse. Two networks are involved—one predicts demand for the seats, up to approximately six months ahead, and takes as input such factors as the day of the week, the time of the flight, and the price of the tickets. The other predicts the no-show rate for each class. There is an interdependency between the classes, since if tickets are available at a large discount there will be correspondingly less demand for the higher-priced seats.

The networks are trained from the airline's historical data, and their output represents the magnitude of each forecast. One of the problems facing the system is that there is no fixed ideal solution, since the optimal marketing and allocation changes as the world changes and different factors come into play. This means that any prediction system must continually adapt to the changing input, a task that is natural for the network but problematical in non-learning systems. This gives the network a distinct advantage and means that it is currently in successful commercial use.

4.13.3   ECG Noise Filtering

An electro-cardiograph (ECG) shows the heartbeat of a patient. However, this heartbeat is not always regular, and the monitoring equipment delivers a signal to a screen that contains so much noise that it can be difficult to see exactly what is going on. The Hecht-Neilsen Neurocomputer Company in America has developed a network that filters out the noise and provides a clean signal. The net has 50 input units, 12 hidden units, and one output unit, and takes as its input 50 time samples of the noisy signal. The magnitude of the output represents the noise-free value at the centre of the time-frame. The net uses the inputs before and after the central value to give contextual information and so assist it in producing the correct output, just as in the NETtalk application. The use of this "past" and "future" information means that the network is always running slightly behind the actual signal being received. For training, 5120 windows on an ECG were digitised from the recorded values of a horse, sampling the input 200 times a second. This set of data was collected carefully so that it was noise-free; noise was then digitally added and the net trained to produce the original version from the noisy one. To give an indication of the success of the approach, the net was tested after 20 passes through the training set on the same input data but with much more noise added. The net produced results that were consistently better than the best adaptive linear filters, and it produced good results even when the noise level was up to 50% of the input signal level.

Another system, developed by Nestor, is able to classify heart-beats, monitoring them when they are normal, and providing an alert if abnormal or potentially dangerous beats occur.

4.13.4   Financial Applications

One of the questions frequently asked of new technology is "can it predict the stock market?", and there have been attempts to apply neural networks to this sort of problem. The most successful pproaches are still likely to be locked away as company secrets, however, and the assumption underlying the question is that it *is* possible to predict the stock market, an assumption that may not be true.

Networks have been developed that have discovered significant patterns in the movement of the markets; notable among these was a program that showed a small set of patterns that frequently occur in the fluctuation of the Japanese yen compared to the U.S. dollar. Other systems have been developed to assist in bond trading, which seem to offer slight improvements over the more conventional computer systems that are already in use; a feature of the financial domain is that a slight improvement in predictive success can be worth a great deal of money. In one comparison, a conventional system predicted the correct move 55% of the time, and was wrong 45% of the time. The network, working on the same inputs as the other system, was undecided 25% of the time. For the remainder, it scored 72% correct, which is an improvement. Equities trading, futures and exchanges are all areas in which networks have been applied, often linking in with other computer prediction systems to provide as accurate a result as possible. One financial system, known as the trader's "assistant", uses a network to extract the significant features from past examples. It then passes them to another system which builds rules around those features. The network is allowed to adapt to the distortion and evolution of the market over a period of months, and so always provides a current set of critical features, which means that the rule base is never out of date.

Another area of financial application is loan scoring; this is the process of deciding to whom it is worth lending money, and how much it is worth lending to them. Delinquency risk assessment, on the other hand, is all about gauging how likely a person is to default on their repayments. Both have seen the successful use of multilayer perceptron networks. The advantage of using neural systems is that they can learn from the many thousands of examples in the company's records, extracting and encoding the relevant features that indicate what is likely to happen. They can not only free human experts from the more mundane jobs to concentrate on more difficult cases, but can also discover important factors that have previously been unnoticed. One particular network system used the information contained in 270,000 previous applications as its training information, using such factors as the applicant's occupation, whether they owned or rented their accommodation, the number of bank accounts they had, and so on. Trained on two passes through the data set, the net was tested on the loan applications for the first half of 1985 since the results of these, in terms of repayment status and profitability, were known. Compared to the company's own approach of using discriminant analysis, the network produced results that would have increased profitability by 7%. In a delinquency risk assessment application, a network of 6,561 nodes was trained on 5 passes over 5000 files, which took about 7 hours—its response to new input took less than one second per file, however. Other companies use networks in the fields of insurance and mortgage underwriting.

4.13.5   Pattern Recognition

Whilst there are many applications of neural networks in diverse fields, the underlying principle upon which they operate is one of pattern recognition, as we have demonstrated earlier. Consequently there are a number of systems that apply themselves directly to the problems of machine vision and object recognition. The Siemens group use networks for industrial scene analysis, as well as being involved in the CMOS design and manufacture of neural chips. Networks have been applied to the problem of aircraft identification, and also to terrain matching for automatic navigation systems. Target identification from sonar traces has also been developed, with some remarkable results. Attempting to distinguish hostile contacts from non-hostile ones, and given only a day of training the network on examples, the net produced 100% correct identification of the target, compared to 93% scored by a Bayesian classifier.

British Rail are currently developing a vision system using a neural network that they hope will assist them to monitor level crossings. The network is designed to produce a high output value if it sees that people are on the crossing, and so act as a safety warning device. There are many difficulties to overcome, though, since the net must produce a consistently high output whenever people are around, whether there is one person or many, whether they are running or walking, adult or child. However, such a network has also to be insensitive to many other things that may appear in its field of view, such as falling leaves, small animals, branches and so on. What is more, it also has to cope with a large variety of lighting conditions from daytime to night, spring to winter, and so the system remains under development.

British Telecom are working, as are many other communications companies, on projects that involve the application of perceptrons, and much of their effort is involved in speech processing, recognition and synthesis. Many firms believe that voice-activated control is much more realistic using neural networks than any other method, and are making great attempts to improve this area of the human-machine interface.

There are many commercial applications of networks in character recognition, ranging from devices that accept handwritten text as input to experimental systems for interpreting hand-drawn diagrams, maps or plans. One of the more widely adopted systems performs signature verification on cheques for the major banks. Due to the high cost in terms of skill and man-hours involved in signature verification, it is usually only done on cheques for large amounts; the majority are simply checked by the cashier glancing at them. Human experts obtain a 50–60% accuracy, a value that is very much dependent on the style of the signature, with flamboyant ones being easier to forge. Given a training set containing 75 examples of the signature, the network achieves an accuracy of between 92 and 98%, in a fraction of the time usually taken. Wider use of the system will soon mean that cheques for smaller and smaller amounts can be automatically verified, saving the bank and its customers a lot of money.

 Summary

- Multilayer perceptron—layers of perceptron-like units.
- Feedforward, supervised learning.
- Uses continuously differentiable thresholding function (usually sigmoid).
- Back-propagation algorithm (generalised delta rule) trains network by passing errors back down the net.
- Three layers of active units can represent any pattern classification.
- Net develops internal representations of the input's structure.
- Repeated presentations of training data required for learning.
- Described by energy landscape.
- Learning process will not always converge.
- Variety of approaches to overcome learning difficulties.
- Radial basis functions separate classes using hyperspheroids and can guarantee convergence.
- Applications varied.

FURTHER READING

1. *Parallel Distributed Processing*, Volume 1. J. L. McClelland & D. E. Rumelhart. MIT Bradford Press, 1986. Referenced in

the previous chapter, it deserves a place here as well since it contains the description of the multilayer perceptron as well as background material.

2. Multi-Variable Functional Interpolation and Adaptive Networks. D. S. Broomhead & D. Lowe. HMSO. RSRE report, April 1988. A paper that shows the mathematics and use of radial basis functions.

3. Parallel networks that learn to pronounce English text. T. J. Sejnowski & C. R. Rosenberg. In *Complex Systems*, 1987, pages 145–168. All about NETtalk.

5

Kohonen Self-Organising Networks

5.1 INTRODUCTION

So far we have looked at algorithms that rely on supervised learning techniques. In this chapter we will explore unsupervised learning methods, and in particular Kohonen's self-organising maps. As we have seen with back propagation techniques, supervised learning relies on an external training response (the desired response of the network) being available for each input from the training class. This technique is very useful, and in some ways relates to the human learning process. However in many applications, it would be more beneficial if we could ask the network to form its own classifications of the training data. To do this we have to make two basic assumptions about the network; the first is that class membership is broadly defined as input patterns that share *common features*, the other is that the network will be able to identify common features across the range of input patterns. Kohonen's self-organising map is one such network that works upon these assumptions, and uses unsupervised learning to modify the internal state of the network to model the features found in the training data. We shall explore this idea fully by looking closely at Kohonen's learning algorithm (and the Grossberg ART network in Chapter 7).

5.1.1 The Self-Organisation Concept

Kohonen—a Professor of the Faculty of Information Sciences, University of Helsinki—has worked steadily in the area of neural networks for many years, long before the current surge of interest erupted in the mid 1980's. He has worked extensively with concepts of associative memory and models for neurobiological activity. His work is characterised by a drive to model the self-organising and adaptive learning features of the brain.

Neurobiologists have long since established that localised areas of the brain, particularly across the cerebral cortex, perform specific functions. Examples might be speech, vision or motion control, each of which can be identified as regions of intense local activity in the brain. More recently, evidence has also been found that suggests even the localised regions may contain further structures which represent the internal mappings of response from sensory organs. A good example is found in the auditory cortex region.

In the auditory cortex it is possible to distinguish a spatial ordering of the neurons which reflects the frequency response of the auditory system. The ordering of the cells within the auditory cortex region trace an almost logarithmic scale of frequency. Low frequencies will generate responses at one end of the cortex region, high frequencies at the opposite extreme. There are arguments for and against the idea of internal neuron mappings of this kind. Those who oppose it argue the case for the so called "Grandmother cell". This idea suggests that individual neurons in the brain are coded to represent a specific concept, for example a specific cell could be responsible for the task of identifying Grandmother. This argument would appear to have little biological justification however—cells in the brain die off at a rather alarming rate for those of us who have passed the first score of our "three score and ten" (typical estimates put the figure at 25000 cells a day). Having an encoding scheme that maps concepts to unique cells cannot be expected to remain reliable with the typical rates of decay of neurons.

The ideas of self-organisation were proposed as early as 1973 by von der Malsburg and followed up in the mid 70's with computer models for self-organisation, by Willshaw and von der Malsburg. Their work was particularly biologically motivated—based on the development of selectively sensitive neurons (i.e. to light intensity and edge orientation) in the visual cortex region. As we discussed earlier in the book, biological learning or adaptation is a chemical process that modifies the effectiveness of the synaptic connections at the input to the neuron cell. There is little doubt that much of the high-level structure is genetically placed and fixed from birth, but this does not account for our continued experience of learning. There is no simple answer to this question—the biological and physiological issues raised are complex. We recently heard the quote "If the brain was simple enough to be understood—we would be too simple to understand it!". Minsky in his book *Society of Mind* elaborates on this complexity and draws the conclusions that the human brain has over 400 specialised architectures, and is equivalent in capacity to about 200 Connection Machines (Model CM–2). (The book is well worth a read if the neurobiological area of this subject interests you.) The outcome of Kohonen's investigations has been the derivation of a neural network learning algorithm based on these concepts of self-organisation, with very plausible extensions to the biological realm.

5.1.2   An Overview

It has been postulated that the brain uses spatial mapping to model complex data structures internally. Kohonen uses this idea to good advantage in his network because it allows him to perform data compression on the vectors to be stored in the network, using a technique known as *vector quantisation*. It also allows the network to store data in such a way that spatial or topological relationships in the training data are maintained and represented in a meaningful way.

Data compression means that multi-dimensional data can be represented in a much lower dimensional space. Much of the cerebral cortex is arranged as a two-dimensional plane of interconnected neurons but it is able to deal with concepts in much higher dimensions. The implementations of Kohonen's algorithm are also predominantly two dimensional. A typical network is shown in figure 5.1. The network shown is a one-layer two-dimensional Kohonen network. The most obvious point to note is that the neurons are not arranged in layers as in the multilayer perceptron (input, hidden, output) but rather on a flat grid. All inputs connect to every node in the network. Feedback is restricted to lateral interconnections to immedi-

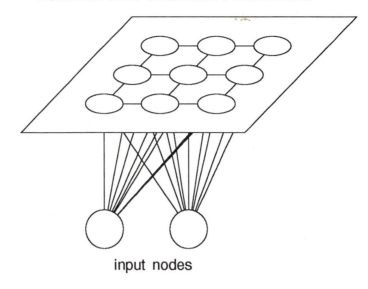

input nodes

Figure 5.1 A Kohonen feature map. Note that there is only one layer of neurons and all inputs are connected to all nodes.

ate neighbouring nodes. Note too that there is no separate output layer—each of the nodes in the grid is itself an output node.

5.2   THE KOHONEN ALGORITHM

The learning algorithm organises the nodes in the grid into local neighbourhoods that act as feature classifiers on the input data. The topographic map is autonomously organised by a cyclic process of comparing input patterns to vectors "stored" at each node. No training response is specified for any training input. Where inputs match the node vectors, that area of the map is selectively optimised to represent an average of the training data for that class. From a randomly organised set of nodes the grid settles into a feature map that has local representation and is self-organised. The algorithm itself is notionally very simple.

Kohonen Network Algorithm

1. Initialise network
Define $w_{ij}(t)\,(0 \le i \le n-1)$ to be the weight from input i to node j at time t. Initialise weights from the n inputs to the nodes to small random values. Set the initial radius of the neighbourhood around node j, $N_j(0)$, to be large. — $N_j = ?$
2. Present input
Present input $x_0(t), x_1(t), x_2(t), \ldots, x_{n-1}(t)$, where $x_i(t)$ is the input to node i at time t.
3. Calculate distances
Compute the distance d_j between the input and each output node j, given by

$$d_j = \sum_{i=0}^{n-1}(x_i(t) - w_{ij}(t))^2$$

4. Select minimum distance
Designate the output node with minimum d_j to be $j*$.
5. Update weights
Update weights for node $j*$ and its neighbours, defined by the neighbourhood size $N_{j*}(t)$. New weights are

$$w_{ij}(t+1) = w_{ij}(t) + \eta(t)(x_i(t) - w_{ij}(t))$$

For j in $N_{j*}(t)$, $0 \le i \le n-1$
The term $\eta(t)$ is a gain term $(0 < \eta(t) < 1)$ that decreases in time, so slowing the weight adaption. Notice that the neighbourhood $N_{j*}(t)$ decreases in size as time goes on, thus localising the area of maximum activity.
6. Repeat by going to 2.

In summary:
- Find the closest matching unit to a training input
- Increase the similarity of this unit, and those in the neighbouring proximity, to the input.

5.2.1   Biological Justification

Is there any biological justification for such a learning rule? As we have seen already, Kohonen has based most of his work on close studies of the topology of the brain's cortex region, and indeed there would appear to be a good deal of biological evidence to support this idea.

We have seen in previous chapters that activation in a nervous cell is propagated to other cells via axon links (which may have an inhibitory or excitatory effect at the input of another cell). However, we have not considered the question of how the axon links are affected by lateral distance from the propagating neuron. A simplified yet plausible model of the effect is illustrated by the Mexican hat function shown in figure 5.2.

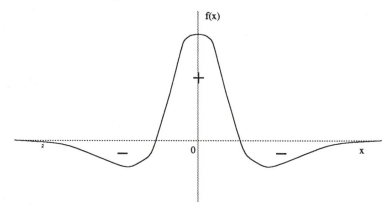

Figure 5.2 The Mexican hat function describes the effect of lateral interconnection.

We can see that cells physically close to the active cell have the strongest links. Those at a certain distance actually switch to inhibitory links. It is this phenomenon to which Kohonen attributes to the development (at least in part) of localised topological mapping in the brain. As we shall see, he has modelled this effect by using only locally interconnected networks and restricting the adaption of weight values to localised "neighbourhoods".

Much of the popularity of this paradigm could be attributed to the fact that it has a very accessible and "natural" feel to it, as we shall hopefully see when we expand the algorithm. Reading through the algorithm, we can see that the learning rule is not complicated. There are no troublesome derivatives to be calculated, as in gradient descent methods. Initially all the connections from the inputs to the nodes are assigned small random weight values. Each node will thus have a unique weight vector, the dimensionality of which is defined by the number of components in the input vector. During the learning cycle, a set of training patterns (a representative subset of the full data set) is shown to the network. The action of the network under the stimuli of these training inputs can be compared to a "winner-take-all" function. A comparison is made between each input pattern, as it is presented, and the weight vectors—the node with the weight vector closest to the input pattern is selected as the "winner". The winning node "claims" the input vector and modifies its own weight vector to align it with the input. The node has now become sensitive to this particular training input and will provide a maximum response from the network if it is applied again after training is completed.

We can see from the algorithm that the nodes in the neighbourhood N_c of the winning node are also modified. The reason for this is that the network is trying to create *regions* that will respond to a spread of values around the training input. The nodes around the winning node are given a similar alignment, and over the course of the training cycle this settles to an "average" representation of that class pattern. As a consequence, vectors that are close spatially to the training values will still be classified correctly even though the network has not seen them before. This demonstrates the generalisation properties of the network.

The two most central issues to adaptive self-organising learning in a Kohonen network are the weight adaption process and the concept of topological neighbourhoods of nodes. Both of these ideas are very different from the neural networks we have discussed so far, so our description of the workings of the Kohonen network will be based around these key themes.

5.3   WEIGHT TRAINING

As we have already mentioned, there is no derivative process involved in adapting the weights for the Kohonen network. Referring to the algorithm again, we can see that the change in the weight value is proportional to the difference between the input vector and the weight vector:

$$w_{ij}(t+1) = w_{ij}(t) + \eta(t)(x_i(t) - w_{ij}(t))$$

where w_{ij} is the ith component of weight vector to node j, for j in the neighbourhood $N_{j*}(t)$ $(0 \leq i \leq n-1)$.

The unit of proportionality is $\eta(t)$, the learning rate coefficient, where $0 < \eta(t) < 1$. This term decreases the adaption rate with time (where by "time" we mean the number of passes through the training set). We can visualise the training cycle as having two stages. The first stage is creating some form of topological ordering on the map of randomly orientated nodes. The training process attempts to cluster the nodes on the topological map to reflect the range of class types found in the training data. This will be a coarse mapping, where the network is discovering how many classes the map must eventually identify, and where they should lie in relation to each other on the map. These are large scale changes to the orientation of the nodes on the map, so the adaption rate is kept high ($\eta > 0.5$) to allow large weight modifications and hopefully settle into an approximate mapping as quickly as possible. Once a stable coarse representation is found, the nodes within the localised regions of the map are fine-tuned to the input training vectors. To achieve this fine-tuning much smaller changes must be made to the weight vectors at each node, so the adaption rate is reduced as training progresses. Typically the fine-tuning stage will take between 100 and 1000 times as many steps as finding the coarse representation, if a low value of η is used.

Each time a new training input is applied to the network the winning node must first be located; this identifies the region of the feature map that will have its weight values updated. The winning node is categorised as the node that has the closest matching weight vector to the input vector, and the metric that is used to measure the similarity of the vectors is the Euclidean distance measure. We discussed this metric earlier in Chapter 2. There are, however, a few subtleties to note in implementing the technique in the Kohonen network. The Euclidean norm of a vector is a measure of its magnitude. However, we are not so much interested in the magnitude of the vectors as in finding out how they are orientated spatially. In other words, we will describe two vectors as being similar if they are pointing in the same direction, regardless of their magnitude. The only way that we can ensure that we are comparing the orientation of two vectors, using the Euclidean measure, is to first make sure *where are* that all the weight vectors are normalised. Normalising a vector *we normalizing in the algo,* reduces it to a *unity length* vector by dividing it by its magnitude— for a set of vectors in Euclidean space this means that each vector will retain its orientation but will be of a fixed length, regardless of its previous magnitude. The comparison of the weight vectors and the input vector will now be concerned only with the orientation, as required. Another useful advantage of normalising the vectors is that it reduces the training time for the network, because it removes one degree of variability in the weight space. Effectively that means that the weight vectors start in an orientation that is closer to the desired state, thus reducing some of the reorientation time during the training cycle.

5.3.1   Initialising the Weights

A note of caution may be inserted at this point concerning the initialisation of the weight vectors. So far we have suggested that on start-up, the network weights should be set to small, normalised random values. However, this is an over-simplification because if the weight vectors are truly randomly spread, the network may suffer non-convergent or very slow training cycles. The reason for this can be explained fairly intuitively. Typically the input training vectors will fall into clusters over a limited region of the pattern space, corresponding to their class (at least it is hoped that they will, else training will be a difficult process). If the weight vectors, stored at the nodes in the network, are randomly spread then the situation could quite easily arise where many of the weight vectors are in a very different orientation to the majority of the training inputs. These nodes will not win any of the best-match comparisons and will remain unused in forming the topological map. The consequence of this is that the neighbourhoods on the feature map will be very sparsely populated with trainable nodes, so much so that there may not be enough usable nodes to adequately separate the classes. This will result in very poor classification performance due to the inability of the feature map to distinguish between the inputs.

The optimum distribution for the initial weights is one that gives the network starting clues as to the number of classes in the training data and the likely orientation that each one will be found, but considering that this is often the very information that we are expecting the network to find for us it is a rather impractical proposition. There are, however, methods to approximate such a distribution.

One method is to initialise all the weights so that they are normal and coincident (i.e. with the same value). The training data is modified so that, in the early stages of the training cycle, the vectors are all lumped together in a similar orientation to the start-up state of the nodes. This gives all the nodes in the feature map the same likelihood of being close to the input vectors, and consequently being included in the coarse representation of the map. As training progresses the inputs are slowly returned to their original orientation, but because the coarse mapping is already defined by this stage, the nodes in the feature map will simply follow the modifications made to the input values. A similar technique adds random noise to the inputs in the early stages of training in an attempt to distribute the vectors over a larger pattern space, and thus utilise more nodes.

It is also possible to attach a threshold value to each node, which "monitors" the degree of success or failure that a node has in being selected as best-match. If a node is regularly being selected, it will temporarily have its threshold raised. This reduces its chance of being voted best-match and allows redundant nodes to be used in forming the features of the map.

The most often used technique, however, and the one quoted by Kohonen, is one that we have already mentioned in passing—that of local neighbourhoods around each node. We will now explain how this maximises the use of all the nodes in the network and promotes topological grouping of nodes.

5.4   NEIGHBOURHOODS

In order to model the Mexican hat function for the lateral spread of activation in interconnected nodes, Kohonen introduces the idea of topological neighbourhoods. This is a dynamically changing boundary that defines how many nodes surrounding the winning node will be affected with weight modifications during the training process. Initially each node in the network will be assigned a large neighbourhood (where "large" can imply every node in the network). When a node is selected as the closest match to an input it will have its weights adapted to tune it to the input signal. However, all the nodes in the neighbourhood will also be adapted by a similar amount. As training progresses the size of the neighbourhood is slowly decreased to a predefined limit. To appreciate how this can force clusters of nodes that are topologically related, consider the sequence of diagrams shown in figure 5.3 that represents the topological forming of the feature clusters during a training session. For clarity, we shall show the formation of just one cluster which is centred about the highlighted node.

In A, the network is shown in its initialised state, with random weight vectors and large neighbourhoods around each node. The arrows within each node can be thought of as a spatial representation of the orientation of each node's weight vector. Training commences as previously described; for each training input the best-match node is found, the weight change is calculated and all the nodes in the neighbourhood are adjusted.

In B we can see the network after many passes through the training set. The highlighted region of the map is beginning to form a specific class orientation based around the highlighted node. The neighbourhood size has also shrunk so that weight modifications now have a smaller field of influence.

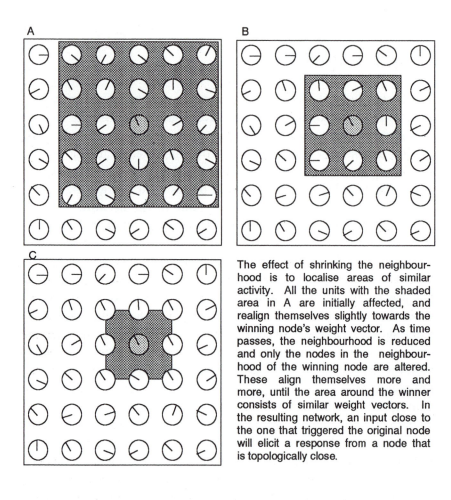

The effect of shrinking the neighbourhood is to localise areas of similar activity. All the units with the shaded area in A are initially affected, and realign themselves slightly towards the winning node's weight vector. As time passes, the neighbourhood is reduced and only the nodes in the neighbourhood of the winning node are altered. These align themselves more and more, until the area around the winner consists of similar weight vectors. In the resulting network, an input close to the one that triggered the original node will elicit a response from a node that is topologically close.

Figure 5.3 Training a localised neighbourhood.

The fully trained network is shown in C. The neighbourhoods have shrunk to a predefined limit of four nodes, and the nodes within the region have all been adapted to represent an average spread of values about the training data for that class.

The training algorithm will produce clusters for all the class types found in the training data. The ordering of the clusters on the map, and the convergence times for training are dependent on the way the training data is presented to the network. Once the network has self-organised the internal representation the clusters on the feature map can be labelled to indicate their class so that the network can be used to classify unknown inputs. Note that the network forms the internal features without supervision, but the classification labelling must be done by hand, once the network has been fully trained.

5.5   REDUCING THE NEIGHBOURHOOD

We have already stressed that the neighbourhood size is reduced with time during the training sequence. But how quickly do we reduce it and to what final size? Unfortunately there are no hard and fast rules for adaptive training algorithms of this nature and some experimentation will be required in individual applications. However Kohonen does stress that his method is not one that is brittle—that is, small changes in system parameters do not reflect gross divergence of training results—and also suggests some rules of thumb as a starting point for intuitive tweaking!

We have explained that the adaption rate must be reduced during the training cycle so that weight changes are made more and more gradual as the map develops. This ensures that clusters form accurate internal representations of the training data as well as causing the network to converge to a solution within a predefined time limit. In typical applications Kohonen suggests that the adaption rate be a linearly decreasing function with the number of passes through the training set.

Training is effected not only by the adaption rate and the rate at which the neighbourhood is reduced, but also by the shape of the neighbourhood boundary. The example we used earlier, in figure 5.3, only discussed the possibility of using a square neighbourhood—however, that is not to say that we cannot define a circular or even a hexagonal region, and these may provide optimal results in some cases. As with the adaptation rate, however, it is preferable to start with neighbourhoods fairly wide initially and allow them to decrease slowly with the number of training passes.

5.5.1   Point Density Functions

For those who prefer a more mathematical definition of what is happening during the training cycle we can explain the clustering phenomenon using probability density functions. A probability density function is a statistical measure that describes the data distribution in the pattern space. For any given point in the pattern space, the probability density function will define a value for the likelihood of finding a vector at that point. Given a pattern space with a known probability density function (i.e. we know how the patterns are spread across the pattern space) it can be shown that the map will order itself such that the point density of the nodes in the feature map will tend to approximate the probability density function of the pattern space (if a representative subset of the data is chosen to train the network). To visualise what this means in practice, consider figure 5.4.

The network is being trained on data from a uniformly distributed pattern space within the two-dimensional outer frame—in other words, the patterns are evenly distributed across the rectangular region. A training set is selected from by choosing independent and random points in the pattern space—the randomness will ensure that a good representation of the total pattern space is provided. The sequence of diagrams represents the state of the weight vectors for various stages of the training cycle. The weight vectors are two dimensional ((X_1, X_2)), and the value of the weight vector defines a point in the weight space. The diagrams are plotted by drawing lines between the points defined by the weights of neighbouring nodes. These plots, then, depict the spatial relationship of the nodes in the weight space (two dimensional in this case). The final state

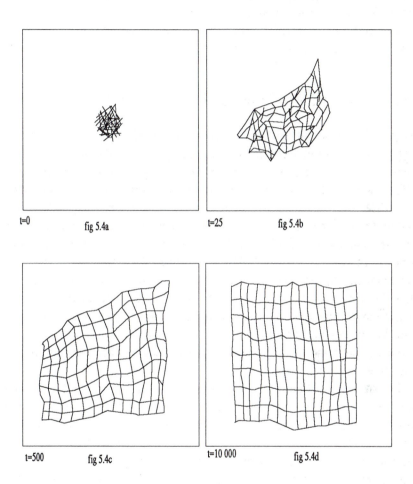

Figure 5.4 Representation of the development of the spatial ordering of the weight vectors.

of the network shown in figure 5.4 shows how the weight vectors have ordered themselves to represent the distribution of the pattern space. The nodes have been optimally ordered to span the pattern space as accurately as possible, given the constraint that there is a limited number of nodes to map the much larger space.

5.6   LEARNING VECTOR QUANTISATION (LVQ)

Despite the fact that the Kohonen network is an unsupervised self-organising learning paradigm, Kohonen does in fact make use of a supervised learning technique. This he describes as learning vector quantisation. This is worth mentioning because it amounts to a method for fine-tuning a trained feature map to optimise its performance in altering circumstances. A typical situation may be that we wish to add new training vectors to improve the performance of individual neighbourhoods within the map.

The way this is achieved is by selecting training vectors (x) with known classification, and presenting them to the network to examine cases of misclassification. Again, a best-match comparison is performed at each node and the winner is noted (n_w). The weight vector of the winning node is then modified according to the following criteria.

For a correctly classified input:

$$n_w(t + 1) = n_w(t) + \eta(t)[x(t) - n_w(t)]$$

For an incorrect classification:

$$n_w(t + 1) = n_w(t) - \eta(t)[x(t) - n_w(t)]$$

The term $\eta(t)$ controls the rate of adaptation, and performs the same function as it did in the learning cycle. The application study that follows shows how this method may be used to add new users to a speech recognition system by optimising the phoneme classifiers on a feature map.

5.7   THE PHONETIC TYPEWRITER

Perhaps one of the best ways of demonstrating the value of an idea is by its successful application. Kohonen has applied his feature map algorithm to the time honoured problem of speech recognition. This is perhaps an ideal application for feature map techniques. The problem is one of classifying phonemes in real time. Why is it so ideally suited to the Kohonen method? The phonemes form a small classification set with class samples showing subtle variations between them. This implies that only a small number of feature detectors need to be formed in the topological map, each of which has many nodes in its neighbourhood tuned over a limited range. That is not to say that phoneme classification is a trivial problem—far from it! Speech recognition is a complex pattern recognition task. Our own human recognition of speech works at several levels of perception. Apart from the fundamental interpretation of the speech waveform, much of the recognition is done at levels applying context, inference, extrapolation, parsing and syntactic rules. We even perform these functions reliably in considerably noisy environments. If you don't believe that statement, think about the cocktail party scenario. We are capable of understanding and holding a conversation in the midst of the general buzz of discussion. We are able to ignore the noise of conversation around us, and yet, if our name is mentioned in conversation elsewhere in the room we are very likely to pick it out (and be worried by it!).

From a signal processing perspective the speech waveform is also ill-defined and complex. Speech phonemes vary in signal strength and shape from speaker to speaker. Even in an individual speaker, phonemes vary in the context of the words that they are formed in, and invariably the spectral signals of the different phonemes overlap for much of their waveform. A great deal of effort has been expended over a sustained period to try to create accurate phoneme classifiers using conventional techniques. The simplicity of the solution we are about to describe serves to show the particular merit of Kohonen's technique.

The driving goal of Kohonen's work was to build a phonetic typewriter—that is a typewriter that could type from dictation. This is perhaps easier in his native tongue of Finnish than it would be in other languages (Finnish being a phonetic language), but it was still quite a complex task. Kohonen has approached the problem applying a mixture of the best of many techniques—he is quick to point out that neural networks are not a universal panacea for all aspects of a data processing problem. The system that he devised is shown schematically in the following figure, figure 5.5.

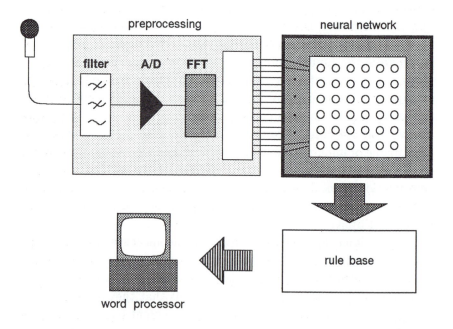

Figure 5.5 A schematic circuit of Kohonen's neural based phonetic typewriter.

The neural network is only dealing with one part of the total task. The system is not totally "neural"—in fact the neural network is only being used in the critical stage of classifying the phonemes. This amounts to vector quantisation of the spectral speech signal. Kohonen was striving for a real-time commercial system—this meant that where conventional computing techniques provide optimal solutions, they were adopted within the system. This is most clearly seen in the front-end signal processing stage.

5.7.1   Front-end Preprocessing

The front-end processing is an essential element to any neural network technique. This point cannot be over-stressed. Any neural network paradigm will perform poorly if given non-representative or inadequate training data. Neural networks do provide a novel method of abstracting feature information into a distributed encoding. They do not, however, by-pass the critical stage in any pattern recognition task of adequately defining the salient and characteristic features of the data. Kohonen's system relies on standard digital signal processing techniques to extract the phoneme spectral data from the voice input. From a microphone input the speech waveform is fed into a 5.3 kHz low pass filter driving a 12-bit A/D converter (at a sampling rate of 13.03 kHz). A 256 point Fast Fourier Transform (FFT) is computed on the digital data from the A/D at 9.83 ms intervals to capture the spectral content of the phonemes. Kohonen uses the FFT technique because it shows the clustering properties of the spectral component better than more conventional coding methods, and thus provides a more useful representation on which the classifier can train. It is also a fast, reliable and well supported technique. The output of the FFT is filtered and made logarithmic before the information is grouped into a fifteen component continuous pattern vector. The information represented in this vector is the instantaneous power in one of fifteen frequency bands ranging from 200 Hz to 5 kHz. Before being applied to the network as input the components have the signal average removed and are then normalised to a constant length. Kohonen also uses a sixteenth vector component to represent other information about the signal. He chose to use this to represent the rms value of the speech signal.

In the preprocessing stage Kohonen has quantised the voice input to a 16-bit feature vector. The feature vector is a short time slice of the speech waveform. These features were used to train the network. It is important to note that the network was not trained on phoneme data, but only the time-sliced speech waveforms. The nodes in the network, however, become sensitised to the phoneme data because the network inputs are centred around phonemes. The network is able to find these phonemes in the training data without them being explicitly defined. The clusters that are formed during training must then be labelled afterwards by hand. This involves presenting isolated phoneme samples to the network and finding the region of maximum response on the topographical map. In Kohonen's experiments 50 samples of each test phoneme were used to label the network after it was trained on voice data. A typical topological feature map is shown in figure 5.6. It shows the trained network with the labelling attached—Kohonen describes it as a phonotopic map.

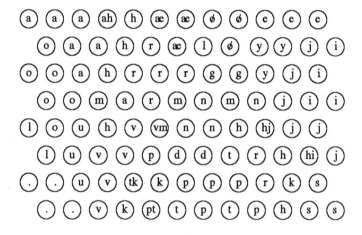

Figure 5.6 A phoneme feature map. Kohonen calls this a "phonotopic map".

The map classifies the more readily defined phonemes—that is, those with relatively stable and predictable speech waveforms. In practice most phonemes have a much longer duration than the sampling rate used of 9.38 ms. The true duration is typically in the order of 40–400 ms. Consequently the classification of a phoneme is made on the basis of several consecutive inputs. Kohonen classifies the phonemes over a number of inputs using simple heuristic rules. One of these relies on the fact that many phonemes have spectra with a unique stationary state by which they can be identified. Alternatively, a sequence of inputs may be monitored—if the phonotopic map's response is constant for a number of consecutive inputs then those inputs correspond to a single phoneme.

It is also possible to visualise the speech waveform as a dynamic trace across the phonotopic map. This is shown in figure 5.7. The steps represent input samples at 9.38 ms intervals. The trace shows the stationary states of the spoken phonemes converging at localised points on the map. Kohonen does not perform classification using these traces, but he does suggest that they provide a new way of visualising phonetic strings, that may be of use in applications such as teaching aids for the deaf.

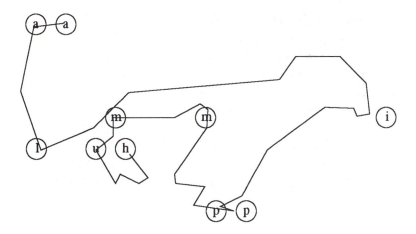

Figure 5.7 The map shows the phonetic trace of the Finnish word "humpilla" across the map.

5.7.2   Auxiliary Maps

The "plosive" phonemes (e.g. b/t/g) have very transient spectra characterised by a high burst of initial energy followed by a period of comparative silence. Kohonen found that the standard phonotopic map did not perform very well at classifying this type of phoneme. His solution was to use auxiliary maps (called transient maps) to classify just the plosive type phonemes. The auxiliary maps were trained on the spectra of the plosive phonemes. The results of such a simple modification to the map was an overall improvement in the recognition accuracy of six to seven per cent.

5.7.3   Post Processing

The last stage of the phonetic typewriter is the translation from the phonetic transcription to orthographic. It is here that the errors from the classification stage must be corrected. The majority of errors are caused by an effect known as coarticulation. Coarticulation is the variation in the pronunciation of a phoneme that is caused by the context of the neighbouring phonemes. To deal with this effect, Kohonen has adopted a rule based system that constructs the correct grammar from the phonetic translation. The rule base is large—typically 15000–20000 rules and deals primarily with context sensitivity of phonemes. It would be impractical to attempt to define rules to account for coarticulation without considering context. The rule base would be prohibitively large if it were to deal with all permutations and it could not cope with the contradictory cases so often found in a language. Kohonen's rule base has been developed from actual example speech data and its correct phonetic transcription. Much like the neural network stage the rules have been derived from example rather than explicitly.

The grammar rule base has been implemented efficiently using hash coding (a software technique for content addressable memory) and operates in near real time—even for a large rule base. The output of the rule base is contextually corrected phonetic strings that can produce orthographic text to drive a word processor environment.

5.7.4   Hardware Implementation

The system has been designed with standard digital hardware. The host computer is an IBM PC/AT with two auxiliary DSP coprocessor boards. Both coprocessors are based on the TMS32010 Digital Signal Processor. One board is responsible for the preprocessing of the speech signal, the other is performing the feature map classification. Post processing is done by the host PC. Even using such standard hardware (much faster DSP's are now available) the recognition system performs at an almost true rate of speech; only a slight pause is required between words.

5.7.5   Performance

The performance figures that follow are quoted by Kohonen from his experiments with the system. Performance figures are always difficult to analyse without understanding the full context of the tests. However, it is worth quoting those for the system as they are fairly indicative of the usefulness of adopting a neural based solution in this type of application.

Correct classification of phonemes from the phonotopic map stage varies between 80 and 90 per cent depending upon speaker and the text. The system accuracy after correction by the grammar rule base increases to between 92 and 97 per cent. This figure is measured by letter accuracy on the orthographic output and is for an unlimited vocabulary.

A performance issue that must also be considered is the flexibility of the system in adapting to new users. Kohonen's system is particularly amenable to the addition of new speakers. They are added using the supervised learning technique, learning vector quantisation, that was described earlier in the chapter. Fine-tuning a map to a new speaker typically requires 100 words and can be completed, according to Kohonen, in ten minutes.

5.7.6   Conclusion

Hopefully, working through this application example has brought two main issues forward. The first is an indication of how self-organising networks may be used in practice. The second is an appreciation of how neural networks may be embedded at the system level. There has been much hysteria concerning the application of neural network techniques and many exaggerated claims for their performance. Neural networks are far from being a universal panacea for all computing situations, but Kohonen's system level approach shows how the strengths of neural techniques (parallelism, generalisation, noise tolerance) may be used in conjunction with conventional techniques to create very powerful computing tools.

 Summary

- Kohonen nets are self-organising, with similar inputs mapped to nearby nodes.
- All the nodes are in one two-dimensional layer.
- "Mexican hat" function of lateral excitation and inhibition.
- Neighbourhood of interactions decreases with time.
- Successfully implemented to produce a phonetic typewriter.

FURTHER READING

1. *Self Organisation and Associative Memory*, third edition. T. Kohonen. Springer-Verlag, 1990. A tutorial introduction to concepts of associative memory and neural networks, with a discussion of self-organisation principles.

2. *Parallel Models of Associative Memory*, second edition. G. E. Hinton & J. A. Anderson. Lawrence Erlbaum Associates, 1989. Collected writings on models of memory and parallel processing—a useful discussion of cognitive and connectionist issues.

3. Self-Organisation of Orientation Sensitive Cells in the Striate Cortex. C. von der Malsburg. In *Kybernetik*, 14, pages 85–100, 1973. An original paper proposing the concept of self-organisation.

4. How patterned neural connections can be set up by self-organisation. D. J. Willshaw & C. von der Malsburg. In *Proc. R. Soc. London*, B. 194, pages 431–445, 1976. Models for the development of self-organised biological networks.

5. Competition and Cooperation in Neural Nets. S. Amari & M. A. Arbib. In *Lecture Notes in Biomath., Vol. 45*, Springer-Verlag (Berlin), 1982.

6

Hopfield Networks

6.1 INTRODUCTION

One of the major contributions to the area of neural networks was made in the early 1980's by John Hopfield, who studied an autoassociative network that has some similarities with the perceptrons studied in earlier chapters, but also some important differences. Hopfield's contribution was not simply the suggestion of a suitable model, but his extensive analysis and study, which has led to his name being associated with the network. He developed the use of an energy function, and related the networks to other physical systems. The Hopfield net consists of a number of nodes, each connected to every other node: it is a *fully-connected* network, and is shown in figure 6.1. An alternative view of the Hopfield net is shown in figure 6.2.

It is also a *symmetrically-weighted* network, since the weights on the link from one node to another are the same in both directions. Each node has, like the single-layer perceptron, a threshold and a step-function, and the nodes calculate the weighted sum of their inputs minus the threshold value, passing that through the step function to determine their output state. The net takes only 2-state inputs—these can be binary $(0, 1)$ or bipolar $(-1, +1)$. However, the bipolar values make the mathematics a little clearer, so we will take the easiest route. What really distinguishes the Hopfield net from the networks discussed earlier is the way in which it produces a solution.

Looking at figure 6.1, or figure 6.2, which show what a fully-connected net is like, there are no obvious input or output connections—each node is the same as any other! This is the major

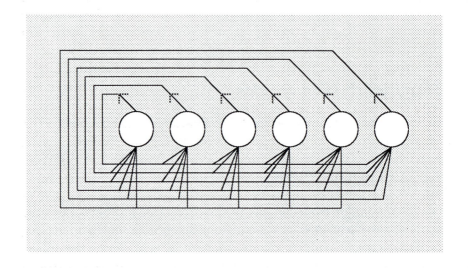

Figure 6.1 The Hopfield network.

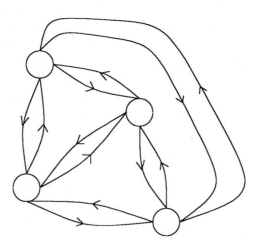

Figure 6.2 The Hopfield network—alternative view.

feature of the Hopfield network, and this difference in architecture means that the network operates in a different way. Inputs to the network are applied to all nodes at once, and consist of a set of starting values, $+1$ or -1. The network is then left alone, and it proceeds to cycle through a succession of states, until it converges on a stable solution, which happens when the values of the nodes no longer alter. The output of the network is taken to be the value of all the nodes when the network has reached a stable, steady state. The reasons behind this behaviour will be outlined in the following sections, but a simple way of visualising the system is to consider that since each node is connected to every other, the value that is on one node affects the value of them all. The initial state represents a lot of different values each trying to affect each other. This is likely to be unstable, since one value may be trying to turn other nodes on, while another is trying to turn them off. As the net moves through a succession of states, it is trying to reach a compromise between all the values in the network, and the final steady state represents the "best compromise" solution that the network can find. In this state, there are as many inputs trying to make a unit turn on as there are trying to make it turn off, so it remains in its stable state.

The operation of the network is radically different from that of a perceptron system, in which inputs are applied and the net produces an output which represents the solution. In the Hopfield net, this first output is taken as the new input, which produces a new output, and so on; the solution occurs when there is no change from cycle to cycle. It is therefore pertinent to ask if the learning procedure is also different. Is there a sensible way to store a set of patterns in a Hopfield net? If so, what is it, and why does it work? In the rest of this chapter, we provide answers to these questions.

6.2 THE HOPFIELD MODEL

 The algorithm governing the operation of the Hopfield net is shown on the following page.

Hopfield Network Algorithm

---

1. Assign connection weights

$$w_{ij} = \begin{cases} \sum_{s=0}^{M-1} x_i^s x_j^s & i \neq j \\ 0 & i = j, 0 \leq i, j \leq M - 1 \end{cases}$$

where w_{ij} is the connection weight between node i and node j, and x_i^s is element i of the exemplar pattern for class s, and is either $+1$ or -1. There are M patterns, from 0 to $M - 1$, in total. The thresholds of the units are zero.

2. Initialise with unknown pattern

$$\mu_i(0) = x_i \qquad 0 \leq i \leq N - 1$$

where $\mu_i(t)$ is the output of node i at time t.

3. Iterate until convergence

$$\mu_i(t + 1) = f_h \left[\sum_{i=0}^{N-1} w_{ij} \mu_j(t) \right] \qquad 0 \leq j \leq N - 1$$

The function f_h is the hard-limiting non-linearity, the step function, as in figure 3.3. Repeat the iteration until the outputs from the nodes remain unchanged.

The weights between the neurons are set using the equation given in the algorithm, from exemplar patterns for all classes. This is the teaching stage of the algorithm, and associates each pattern with itself. The recognition stage occurs when the output of the net is forced to match that of an imposed unknown pattern at time zero. The net is then allowed to iterate freely in discrete time steps, until it reaches a stable situation when the output remains unchanged; the net thus converges on the solution. The autoassociation of patterns should mean that the presentation of a corrupt input pattern will result in the reproduction of the perfect pattern as the output— the network therefore acts as a content-addressable memory. (Refer to Chapter 8 for further details on associative memories and their general properties.)

The operation of the Hopfield network can be summarised as

- initialise the network
- input unknown pattern
- iterate to convergence.

6.3   THE ENERGY LANDSCAPE

The Hopfield net can best be understood in terms of the now ubiquitous energy landscape. We have seen how successful it is in describing the behaviour of perceptrons, since it provides a visual analogy that allows us to form an intuitive view of what is happening. The same is true for a Hopfield network. The energy landscape has hollows that represent the patterns stored in the network. An unknown input pattern represents a particular point in the energy landscape, and as the network iterates its way to a solution, the point moves through the landscape towards one of the hollows. These basins of attraction represent the stable states of the network. The solution from the net occurs when the point moves into the lowest region of the basin; from there, everywhere else in the close vicinity is uphill, and so it will stay where it is. This is directly analogous to the three-dimensional case where a ball placed on a landscape of valleys and hillsides will move down towards the nearest hollow, settling into a stable state that doesn't alter with time when it reaches the bottom.

We can express this in more detail if we look at it mathematically. The energy function for the perceptron was $E = \frac{1}{2}\sum(t_{pj} - o_i)^2$, but this depends on knowledge of the required output as well as the actual output of the net. For the Hopfield net, which steps its way towards a solution, the required intermediate steps aren't known, and we therefore need something more applicable to this architecture. However, it would be sensible to retain some of the features of the perceptron energy function: it should be large for large errors, and small for small errors. The weight values in the network must affect the energy, as must the actual patterns presented, so the energy function must reflect these requirements.

We can identify a suitable energy function for the Hopfield net as

$$E = -\frac{1}{2}\sum_i\sum_{j\neq i} w_{ij}x_ix_j + \sum_i x_iT_i \qquad (6.1)$$

where w_{ij} represents the weight between node i and node j of the network, and x_i represents the output from node i. The threshold value of node i is represented as T_i. As the output is fed back into the net, the outputs at any one time represent the next set of inputs, and so both the weights and the inputs are explicitly represented as required. The weights in the network contain the pattern information, and so all the patterns are included in this energy function. Nodes are not connected directly to themselves, and so the terms w_{ii} are zero. Since the connections are symmetric, $w_{ij} = w_{ji}$.

Having defined our error function we can now answer the questions posed earlier about storing and recalling patterns. If we make our patterns occupy the low points in the energy landscape, then we can perform gradient descent on the energy function in order to end up in one of these minima, which will represent our solution.

6.3.1   Storing Patterns

In order to store a pattern, we need to minimise the value of the energy function for that particular pattern so that it occupies a minimum point in the energy landscape. However, we also want to leave any previously stored patterns in their hollows, so that adding new patterns doesn't destroy all the previous information. The weight matrix contains the information about the stored patterns, so we want to try to find an expression for the weight values that will produce a minimum in the energy function.

Considering this in terms of the energy function, we want to minimise

$$E = -\frac{1}{2}\sum_i\sum_{j\neq i} w_{ij}x_ix_j + \sum_i x_iT_i$$

for a particular pattern s that has a set of input elements $(x_0, x_1, \ldots, x_{n-1})$.

We want each term to be negative, and so we require $\sum_i x_i T_i$ to be negative. This can be achieved by setting T_i to the opposite sign of x_i for a particular pattern. However, a different pattern will have different values of x_i and then the threshold term may well increase the value of E. In order to avoid this, the best that we can do is to set the threshold to zero, which will not decrease or increase the value of the energy function for any of the patterns.

We write x_i^s to mean element i of input pattern s, which can be either $+1$ or -1. Now, w_{ij} is the weight between nodes i and j as before, and contains the pattern information from all the taught patterns. This means that we can split the weight matrix into two parts, one which represents the effects of all the patterns *except* the sth one, denoted by w'_{ij}, and a second which is the contribution made by the sth pattern alone, shown as w_{ij}^s. This means that we can rewrite the energy function in two parts

$$
\begin{aligned}
E &= -\frac{1}{2} \sum_i \sum_{j \neq i} w'_{ij} x_i x_j \\
&\quad -\frac{1}{2} \sum_i \sum_{j \neq i} w_{ij}^s x_i^s x_j^s \\
&= E_{\text{all except s}} + E_s
\end{aligned}
\tag{6.2}
$$

where we have separated the contribution made to the energy function from the sth pattern. This can be thought of as viewing the energy as a "signal" plus a "noise" term; the "signal" is the energy due to the pattern s, whilst the "noise" is due to the contributions from all the other patterns.

Storing this pattern corresponds to making the energy function as small as possible. The first term corresponds to the "noise", and we cannot do much to alter this, but we can reduce the contribution made by the second, "signal" term. In other words, to store pattern s, we want to minimise the contribution to the energy function from the sth energy term, and so make

$$E_s = -\frac{1}{2} \sum_i \sum_{j \neq i} w_{ij}^s x_i^s x_j^s \qquad (6.3)$$

as small as possible.

This corresponds to making

$$\sum_i \sum_{j \neq i} w_{ij}^s x_i^s x_j^s$$

as large as possible, due to the minus sign in (6.3).

Now, the elements in x_i are either $+1$ or -1; however, x_i^2 is *always* positive, so if we make the energy term dependent on $x_i^2 x_j^2$ it will always be positive, and so the sum will be as large as possible.

We can do this most simply by equating

$$\sum_i \sum_{j \neq i} w_{ij}^s x_i x_j = \sum_i \sum_{j \neq i} x_i^2 x_j^2$$

and noticing that all we have to do is to make the weight term

$$w_{ij}^s = x_i x_j$$

This means that we have our result; setting the values of the weights $w_{ij}^s = x_i x_j$ minimises the energy function for pattern s. In order to calculate the weight values for all the patterns, we sum this equation over all patterns to get an expression for the total weight set between nodes as

$$w_{ij} = \sum_s w_{ij}^s = \sum_s x_i^s x_j^s$$

Comparing this to step 1 of the algorithm, we see that they are identical, and we now know that step 1 really does store all the initial patterns in the network.

Referring to equation (6.3), altering the w_{ij} each time will alter the value of $E_{\text{all except}s}$ somewhat, so adding patterns does disrupt the previous storage to some extent, but this is unavoidable.

The Hopfield net has no iterative learning algorithm as such; patterns are simply stored by lowering their energies. The network has no hidden units, and so is unable to encode the data.

6.3.2   Recall

Having stored our patterns in the net, we now need to be able to recall them. This can be accomplished if we perform gradient descent on our energy function, so we need a method to do this. Considering our energy function in (6.1), we need to calculate the contribution that a particular node's value makes to the energy, and then we can cycle around the net, reducing each node's contribution until the energy value is at a minimum.

We can express the energy function in two parts, splitting off the contribution made by the kth node.

$$E = -\frac{1}{2}\sum_{i\neq k}\sum_{j\neq k} w_{ij}x_i x_j + \sum_{i\neq k} x_i T_i$$
$$-\frac{1}{2}x_k \sum_j x_j w_{kj} - \frac{1}{2}x_k \sum_i x_i w_{ik} + x_k T_k \qquad (6.4)$$

The kth neuron changes output state from x_{k1} to x_{k2}. The difference in energy $\Delta E = E_2 - E_1$ caused by the state change $\Delta x_k = x_{k2} - x_{k1}$ is given by evaluating equation (6.4) for x_{k2} and x_{k1}, then subtracting, and can be written as

$$\Delta E = -\frac{1}{2}\left[\Delta x_k \sum_j x_j w_{kj} + \Delta x_k \sum_i x_i w_{ik}\right] + \Delta x_k T_k \qquad (6.5)$$

The first two terms in (6.4) are unaffected by the alteration of neuron k, and so remain unchanged and cancel out. Since the matrix w_{ij} is symmetric, we can interchange the indices and simplify the expression to

$$\Delta E = -\Delta x_k \left[\sum_j x_j w_{kj} - T_k\right] \qquad (6.6)$$

$\sum_j x_j w_{kj}$ is the weighted sum of the inputs to node k, and T_k is the threshold of unit k. Now, the threshold of every node was set to zero in the storage phase, in order to ensure that the patterns occupied the minima in the energy function. Remembering that the node's output is either a $+1$ or a -1, decreasing ΔE_k will mean outputting a $+1$ if the weighted sum is greater than zero, and -1 if it is less than zero, since this will always serve to reduce the value of ΔE_k. If we compare this to the update function for nodes in a Hopfield network, given by

$$\sum_{j \neq k} w_{ij}x_i \begin{cases} > 0 & x_i \to +1 \\ = 0 & \text{remain in previous state} \\ < 0 & x_i \to -1 \end{cases}$$

we can see that the update function performs this operation, and so implements gradient descent in E. This allows us to recall our patterns from the net by cycling through a succession of states, each of which has a lower energy (or, if the weighted sum is equal to the threshold, an equal energy) than the previous one. This relaxation into lower energy states continues until a steady state of low energy is reached, when the net has found its way into a minimum and so produced the pattern.

There are two subtly different methods of actually performing the update, which produce slightly different behaviour in the network. The update can be carried out on all nodes simultaneously, where the values in the network are temporarily frozen and all the nodes then compute their next state. This new state represents one update across the entire network, and the following state is then computed. This operation is known as *synchronous* updating. The alternative approach, called *asynchronous* updating, occurs when a node is chosen at random and updates its output according to the input it is receiving. This process is then repeated. The main difference between the methods is that in the case of asynchronous updating, the change in output of one node affects the state of the system and can therefore affect the next node's change. This means that the order in which the nodes are updated affects the behaviour of the network to some extent. The effects are evident in the recall stage, since the random nature of the choice of the next node to be updated alters the sequence of patterns that the network evolves through. With synchronous updating, all nodes are updated together and so the intermediate patterns do not alter. The asynchronous updating therefore adds some uncertainty, or *non-determinism* into the path that will be taken from the input to the final steady state. However, both methods share the same general characteristics of the network, and the use of synchronous or asynchronous updating is rarely an important factor.

The assumptions made in the Hopfield network of a symmetric, zero-diagonal weight matrix are central to its operation. Even slight deviations from this symmetry can give rise to networks that are unstable and do not settle into any final state. One of the current research areas is the investigation of different connectivities and the effects that these have on the behaviour of the network. Hopfield himself has extended the model in a different direction, showing that a smooth function like the sigmoid can be used, with similar results to that of the step function.

6.3.3   An Example

Figure 6.3 shows a set of patterns that were used to train a Hopfield network. Figure 6.4 shows how the network operates. It is presented

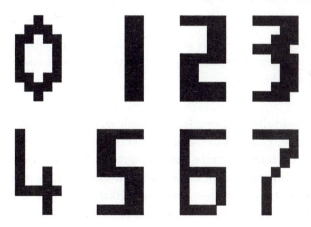

Figure 6.3 The training set for the Hopfield network.

with a corrupted pattern as input, and proceeds to cycle through a series of intermediate states, slowly recovering the correct solution as shown. Each pattern in the sequence has a lower energy than the one before it, and these patterns keep evolving until the network reaches a minimum in the energy function, at which stage the outputs are unaltered from cycle to cycle and the net has produced its solution.

Figure 6.4 The network is presented with a corrupt input pattern, and the sequence shows how it net cycles through successive states until it has recovered a stable result.

Experimentally, the overlap between stored patterns, as we mentioned above, causes interference effects and errors occur in the recovery of patterns if more than about $0.15N$ patterns are stored, where N is the number of nodes in the network. This means that for a network with 100 nodes, errors are likely to occur if the number of patterns stored exceeds 15. These error states are stable outputs from the network that do not correspond to any taught patterns—in terms of the energy landscape, there has been sufficient interference between patterns to form intermediate local minima states that were not taught to the network, but which the network thinks are perfectly acceptable solutions. Such states are known as *metastable* states.

6.4   THE BOLTZMANN MACHINE

The Hopfield net converges to local minima which may not give the optimal solution. We need a method that allows us to escape from these local hollows and move into some deeper well that represents a better result. If the solution to the inputs is represented as a small ball on the energy landscape, it is easy to imagine that giving this ball some intrinsic energy (thermal energy) will allow it to randomly move about in the potential wells and probably escape from local minima. This "shaking" of the nominally stable situation needs to be done carefully however, as violent shaking is as likely to move the solution away from a stable point as towards it. The best method is to provide a lot of energy at first, and slowly reduce the amount as the network works its way towards a global solution.

This idea is similar to that in metallurgy, where the low energy state of a metal is produced by melting it, then slowly reducing its temperature. This annealing of a metal ensures that it reaches a stable, low energy configuration.

Thermal noise is added to the network; to begin with, high temperatures are simulated resulting in a lot of thermal noise, then "temperature" is slowly lowered so that the amount of thermal noise decreases. This is achieved by using a similar structure and learning algorithm to the Hopfield net, coupled with a *probabilistic* update rule. This network is called a Boltzmann machine. Each node in the network calculates which state it should switch into to reduce the energy, as before, but instead of just switching, it changes to that state depending on the value of the probability function. This means that sometimes the network doesn't switch into a lower energy state, but allows jumps to be made into higher energy states, and it is this feature that allows local minima to be escaped. The probability function is chosen so that if the unit will achieve a great reduction in the overall energy by changing its state, then it will probably be allowed to change, but if there isn't a great deal to be gained either way, the likelihood of it changing is much more uncertain. It also has a parameter to vary its "temperature"—at high temperatures, jumps to higher energy states are much more likely to occur than at lower temperatures. As the temperature is lowered, the probability of assuming the correct low energy state approaches one, and the network is said to have reached *thermal equilibrium.*

We can express this mathematically as follows. Each unit in the network computes an energy gap, given by

$$\Delta E_k = \sum_i w_{ki} s_i - \theta_k \tag{6.7}$$

and switches into the state that is of lower energy according to the probabilistic update rule, i.e. with probability

$$p_k = \frac{1}{1 + e^{-\Delta E_k/T}} \tag{6.8}$$

The network can settle into one of a large number of global energy states, the distribution of which is given by the *Boltzmann* distribution. If we let P_α be the probability of the network settling into some global energy state of energy E_α, then the Boltzmann distribution has the form

$$P_\alpha = k e^{-E_\alpha/T}$$

i.e. it is dependent on the energy of the state and the temperature of the system. Calling P_β be the probability of a state with energy E_β, we can write

$$
\begin{aligned}
\frac{P_\alpha}{P_\beta} &= \frac{e^{-E_\alpha/T}}{e^{-E_\beta/T}} \\
&= e^{-(E_\alpha - E_\beta)/T} \tag{6.9}
\end{aligned}
$$

The network is allowed to settle into thermal equilibrium, when the probabilities of states no longer alter, and are dependent on their energy. If E_α is a lower energy state than E_β, then we can see that

$$
\begin{aligned}
E_\alpha &< E_\beta \\
e^{-(E_\alpha - E_\beta)/T} &> 1 \\
\text{therefore } P_\alpha/P_\beta &> 1 \\
\text{so } P_\alpha &> P_\beta
\end{aligned}
$$

This means that as the network approaches thermal equilibrium, lower energy states are more probable, dependent only on their relative energy.

At high temperatures, the net reaches equilibrium quickly, but good global energy states are not much more likely to occur than poor ones. Reducing the temperature while the network is running is called *simulated annealing*, and allows the system to reach low temperature equilibrium in the quickest way possible. The high temperatures allow local minima states to be escaped via higher energy states, but also allow transitions from lower minima to higher ones with almost equal probability. As the temperature is lowered however, the probability of escaping from a higher energy minima to a lower one falls, but the probability of travelling in the reverse direction falls even faster, and so more low energy states are reached. Eventually the system settles down at low temperature in thermal equilibrium. This means that it is the output probabilities of the states that become constant, not the values of the states themselves. The effect of the temperature on the probabilistic function that governs the chance of an unit changing state is shown in figure 6.5. Notice that in high temperature situations, the probability of changing into a higher energy state for any particular input value is greater than for lower temperature situations.

This description of simulated annealing is an oversimplification; since the energy landscape is a highly dimensional space, the energy barrier between states is usually massively degenerate. This means that there are many ways of passing from one state to another, which increase exponentially with the amount of thermal noise added to the system. With such a large number of paths along which to escape, it is even more likely that the system will move into the lower energy state.

The temperature alteration is achieved by adjusting the steepness of the sigmoid function, which effectively determines the probability that a unit will actually go into its natural, or non-noisy, state. If the unit exceeds the threshold by a large amount then it will always attain value 1, whilst if it is far enough below the threshold, then it will always have value 0. Just above threshold, the probability

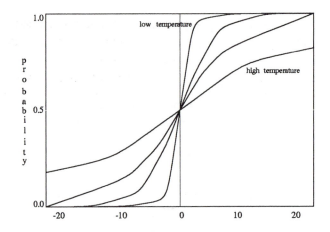

Figure 6.5 The effect of temperature on the transition probability function $1/[1 + \exp(-\Delta E_k/T)]$. The probability of a transition to a higher energy state is greater at higher temperatures than it is at lower ones.

of becoming 1 is greater than 1/2, and just below threshold, the probability of turning off is greater than 1/2. Decreasing the temperature decreases the probability that a unit will have its natural state altered. The function described above follows the Boltzmann distribution, just as in statistical mechanics.

The rate at which the temperature is decreased is important, since this affects the opportunities that the network has to develop a globally optimal solution. If the temperature is lowered too quickly, the net does not have enough opportunity to escape from local minima and so a good solution is not reached. Conversely, if the temperature is lowered very slowly, the network can escape from local minima but will take a long while to converge to a final solution. Examination of the behaviour of the network can help to alleviate this problem to some extent however. At high temperatures, the net moves into high energy states easily, and the overall energy of the system is high. At the other extreme, low temperatures mean that transitions to higher energy states are extremely rare, and the net will tend to stay in its current state of relatively low energy. However, the transition between these two states is not a gentle one, since there is a period during the lowering of the temperature when the transitions from higher to lower energy minima occur much more often than transitions in the opposite direction, from low energy minima to high. It is during this period that the overall energy of the network decreases most rapidly, and so the time spent in this transition period should be as long as possible.

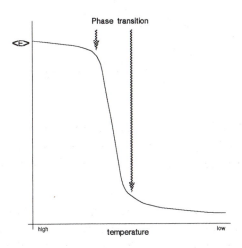

Figure 6.6 Graph of average energy of the network (y-axis) plotted against temperature (x-axis). The phase transition is clearly shown, where the mean energy of the system falls very quickly for a small reduction in temperature.

This behaviour is akin to the phase transitions encountered in substances as they cool and change from one state to another—there is a critical temperature (the melting or boiling point!) at which the state of the system suddenly changes from a high overall energy to a much lower one. The phase transition for a Boltzmann machine is sketched in figure 6.6. Fastest convergence to a good global minimum will occur if the temperature is decreased in such a way as to spend most of the time in the phase transition part of the graph. However, actually determining the phase transition point in practice is difficult.

6.4.1   Learning in Boltzmann Machines

Learning occurs in two phases in Boltzmann machines. The network is fully connected, but an arbitrary choice is made as to which units are to be input units and which are to be output units. In the first phase, the input and output units are clamped to their correct values. The net is then allowed to cycle through its states, with the temperature being gradually lowered until the hidden units reach thermal equilibrium. Weights that connect two units that are both on are then incremented. In the second phase, only the inputs are clamped to their correct values, with the hidden and output units left free. The net runs as before until it reaches thermal equilibrium, and then weights between any two units that are on are decremented. The first phase reinforces connections that lead from the input to the output, whilst the second "unlearns" poor associations.

In a Boltzmann machine, the deepest global minima are usually reached since the system can escape from local minima by allowing jumps to intermediate higher energy states, and the probability that the system settles into these minima is dependent only on the energy of the state, as shown by equation (6.9). In other words, the system is most likely to occupy the best minimum. This fact allows us to observe that this is a recall procedure if all our patterns occupy global minima; therefore, if we can find a way to store the patterns in the global minimum states then we have a ready-made recall method. Alternatively, since there is a direct relationship between the probability of a state occurring and its global energy, we can store probability distributions in our network, by making the energy of a particular state proportional to the probability of it occurring. This gives us a direct representation of probability in a system. The learning procedure for Boltzmann machines which achieves this is given below.

First, we choose arbitrary units in the network to be input units and output units, with the remainder assuming the role of "hidden" units, as in figure 6.7.

The distinction between layers is not as clear as in multilayer perceptrons; the Boltzmann net is fully interconnected with the output units connected back up to the input units and the hidden units.

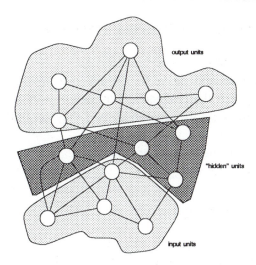

output units

"hidden" units

input units

Figure 6.7 Visualisation of the division of the fully-connected net into "layers" For clarity, not all the connections between units have been shown.

 As we have stated, the learning procedure is in two phases, an *incremental* stage and a *decremental* stage.

Boltzmann Machine Learning Algorithm

Phase 1—incremental
1. Clamp the input and output units to their correct values.
2. Let the net cycle through its states. Calculate the energy of a state

$$\Delta E_k = \sum_i w_{ki} s_i - \theta_k \qquad 0 \le i \le N - 1$$

then switch to lower energy state with probability p_k where

$$p_k = \frac{1}{1 + e^{-\Delta E_k/T}}$$

Reduce T until output is stable.

3. Increment the weight between two units if they are both on.

Phase 2—decremental
1. Clamp the input *only*, leave the output and hidden units free.
2. Let the net reach thermal equilibrium again—run as in Phase 1.
3. Decrement the weights between units if they are both on.

Continue this until the weights are stable.

6.4.2   Why does this work?

We can see how this algorithm achieves learning by considering the behaviour of the weights in the system as they are altered. With a forced output, the weights between "on" units are incremented in phase 1. Notice that this is Hebbian learning—incrementing weights between active units. If the net produces the same output in phase 2, showing it has learnt the correct response, then the same weights will be decremented, and the two phases will cancel each other out. However, if the output is not matched, then some of these weights will be left on, whilst others will be turned off. After a period of time, only the weights between units that produce the correct output will have been left on.

6.4.3   Mean Field Theory

One of the problems with the simulated annealing process is that the probability of switching into a state is calculated by summing the weighted outputs minus the threshold (equations (6.7) and (6.8)) of all the other units in the network. Because these units are also changing their output over time, we ought to calculate the probability based on the *average* output of the other units, and this takes time to compute. We can simplify the problem by replacing the binary state of a unit by a real number which represents the probability of that unit being in the on state, and use this to estimate its average effect on the unit in question. This is similar to "mean field theory" in physics, where the average effect of different fields acting on a particle is approximated by the effect of the average of the different fields (figure 6.8).

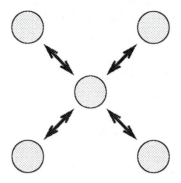

Figure 6.8 The mean field approximation: each unit feels the effects of the average of the other units.

Expressed mathematically, and using

$$G(x) = 1/(1 + e^{-x})$$

and $\langle x \rangle$ to represent the mean value of x, our correct expression can be written

$$p_i = \left\langle G\left(\frac{1}{T}\sum_j s_j w_{ij}\right)\right\rangle \tag{6.10}$$

where s_j are stochastic states; this equation represents the average effect of the varying states.

The approximation is written as

$$p_i = G\left(\frac{1}{T}\sum_j p_j w_{ij}\right) \tag{6.11}$$

where

$$p_j = \langle s_j \rangle \tag{6.12}$$

This represents the effect of the average states.

In using the mean field approximation we introduce errors into the system, since we cannot represent the average of the s_j states accurately. However, we avoid the sampling error at equilibrium since we have represented the units' output probabilities directly. These errors mean that the Boltzmann learning procedure is no longer strictly correct, but the system still works, and it does so *much* faster. In summary, the mean field net therefore approximates the Boltzmann machine but operates much more quickly.

6.4.4   Spin Glasses

Hopfield made a great breakthrough in the understanding of the behaviour of Hopfield and Boltzmann nets by demonstrating that their behaviour could be expressed in terms of the energy function, and that the energy function itself was similar to the energy function encountered in the world of spin glass theory in physics.

Spin glasses are disordered, frustrated magnets. That is, they are materials comprised of particles that each have a particular "spin", which makes them want to be aligned in a common direction. However, there are usually additional forces trying to align the particles differently, such as the presence of an external field or localised effects due to surrounding particles; these competing ordering instructions are what is meant by "frustrated". The "disordered" refers to the fact that there are quenched random constraints on the particles, due perhaps to the lattice in which they find themselves.

The behaviour of these systems has been studied in detail by physicists, and the form of the energy function, the structure of the energy space, and the stable states of the system are known.

We can see how the two systems are analogous by considering the form of their energy functions. Although we will go no further than to demonstrate the mapping between the two, this is important since it shows that the techniques of statistical mechanics can be applied in the analysis of highly-connected networks.

The two states of the model neuron, "on" and "off", can be represented by the states $x_i = +1$ or -1 which we can interpret as "spin up" or "spin down" states. The behaviour of the system is described by the energy function given before (equation (6.1)), quoted again for convenience

$$E = -\frac{1}{2}\sum_i \sum_{j \neq i} w_{ij} x_i x_j + \sum_i x_i T_i$$

The weights w_{ij} can be positive or negative, and the total input to a neuron is given, like before, as

$$U_i = \sum_j w_{ij} x_j \qquad (6.13)$$

The output of the neuron is given by a probability function dependent on the sum of weighted inputs minus some threshold, passed through a non-linear function as usual:

$$p_i = \Theta(U_i - T_i)$$

We can consider the idealised case where Θ is the Heaviside function, and the probability that the state is simply equal to the value of that function is 1, as in the Hopfield net. Therefore the output at the next time interval $(t+1)$ is given by

$$x_{i(t+1)} = \text{sign}(U_{i(t)} - T_i) \qquad (6.14)$$

Stability requires that the outputs be equal to the inputs, and so $x_{i(t+1)} = x_{i(t)} = $ constant. Noting equation (6.13), (6.14) can be written as

$$x_i = \text{sign}\left(\sum_j w_{ij} x_j - T_i\right) \qquad (6.15)$$

If the output from the network is stable, then this equation will hold.

Compare this to the situation in spin glass theory, where the particles' behaviour is governed by an energy function of the form

$$H = -\sum_{ij} w_{ij} \sigma_i \sigma_j - \sum_i h_i \sigma_i \qquad (6.16)$$

The first term corresponds to the interactions between pairs of particles, and the second to locally pervasive effects on single particles. The matrix w_{ij} is symmetric, so $w_{ij} = w_{ji}$. The condition for stability against single spin flips from $\sigma_i \rightarrow -\sigma_i$, which will increase the energy of the system, is

$$\sigma_i = \text{sign} \left[\sum_j w_{ij}\sigma_j - h_i \right] \qquad (6.17)$$

Comparison of equations (6.1) and (6.16) show that a similar energy function governs the operation of the two systems, and what is more, the conditions for stability are comparable as well. Equations (6.17) and (6.15) are identical for symmetric weights and an appropriate choice of h_i to match w_{ij} and T_i.

The use of spin glass theory in the analysis of Hopfield networks has been very successful, and some of the more important results are outlined below. There are expected to be two classes of spurious states for finite values of the storage density $\alpha = P/N$, where P is the number of stored patterns and N is the number of nodes. These are metastable states other than the stored pattern states, and mixture states which overlap with several of the stored prototypes. In addition, there are "spin glass" states, which bear little relation to the stored states and can therefore be considered as spurious for memory retrieval. Pattern retrieval is possible up to about $\alpha = 0.15$ with little error, whilst above this there is a sharp collapse in the retrieval ability of the network. The effect of temperature on the system acts like noise, and for low values can smooth the energy surface and eliminate metastable states, but for high values no retrieval solutions, only spin glass ones, are found.

6.5   CONSTRAINT SATISFACTION

The Boltzmann machine produces solutions that are equivalent to minima in the energy function. We can use the Boltzmann machine, like the Hopfield net, as a content-addressable memory, by ensuring that we make the patterns stored occupy the minima in this energy function. This corresponds to finding an optimal output for the given energy function, since the network converges to some minimum value. It is therefore possible that the same network design can optimise other problems. If we can express the constraints that we want to satisfy in terms of a suitable energy function, the network will produce a solution to that function that minimises the energy. This means that we have to construct our energy function so that it represents the constraints that we wish to minimise. For example, if we wish to minimise the cost of transporting goods, and the cost is proportional to the distance the goods have to be moved, the energy function will have to be large when the distances involved are large, and small when the journeys are short. Minimising this will then correspond to minimising the transport costs.

6.5.1 The Travelling Salesman Problem

The Boltzmann machine can be used for much more complex constraint satisfaction involving a number of possibly conflicting requirements. One of the most interesting problems of this nature is known as the "travelling salesman problem" (TSP), and has been studied by many different people using different techniques. It is widely used as a test problem, and is to constraint satisfaction problems what the XOR problem is to pattern classifiers. The TSP problem is this: imagine you are a travelling salesman for a company. You have to visit all the cities in your area, returning home when finished, but you don't want to visit any city more than once. The cities are different distances apart, and the problem you face is to decide the shortest route for you to take.

The *best* solution to a TSP is very difficult to find, and the time taken to solve it grows exponentially as the number of cities in the tour increases. For this reason, any "good" solution will do. It is a constraint satisfaction problem, the constraints being that each city must be visited once and only once, and that the distance travelled between cities must be as short as possible. If such an energy function is constructed, then minimising that function corresponds to producing a solution that optimises the constraints. In order to solve the TSP problem, it has to be cast into a form which the network can represent. Since the solution is a list of cities to be visited in a particular order, we need an approach which allows us to specify both the city and the position in which it is visited. If there are n cities, each can occur in one of n positions. If we assume that the city to be visited is represented by a neuron with an on state, then, as we want to visit only one city at a time, all the neurons representing the other cities must be off. For n cities, we need n neurons to represent this. As an example, in a 4 city problem, if city A is to be visited first, then we want the first neuron to have its output set to 1, with all the others at 0

$$1\ 0\ 0\ 0$$

Since we need n neurons to represent the position in the tour of one city, it follows that we need n neurons to represent the positions of the n cities. Therefore, the representation that we can use for the TSP problem is a square matrix of n by n nodes, in which the cities are represented along one side, and the possible positions along the other. An example of this for a 4 city tour is shown below.

City \ Position	1	2	3	4
A	1	0	0	0
B	0	0	1	0
C	0	0	0	1
D	0	1	0	0

In this example, city A is visited first, then D, B and lastly C. We need to construct our energy function so that minimum states correspond to good solutions. It needs to produce short paths, but must also represent valid tours. A valid tour occurs when each city is visited only once, which corresponds to there being only one term set to 1 in the rows of the matrix, and that each city is visited, which means that there must be one, but only one, term set to 1 in the matrix columns. Another constraint which helps to promote valid tours is that there should be no more than n 1's in the matrix as a whole. If we write V_{Xi} to represent the network's outputs, then the X subscript represents a city name and the i subscript the position it appears in the tour. We can write an energy function for these conditions as

$$E = A\sum_{X}\sum_{i}\sum_{j\neq i}V_{Xi}V_{Xj}+B\sum_{i}\sum_{X}\sum_{X\neq Y}V_{Xi}V_{Yi}+C\left(\sum_{X}\sum_{i}V_{Xi}-n\right)^{2}$$

The first term is zero if and only if each city row X contains only one 1. The second term is zero if and only if each position in the tour column contains only one entry set to one, whilst the third term is zero if and only if there are n entries in the matrix as a whole. These terms therefore favour states that correspond to valid tours for the TSP.

We need to add another term to this energy function in order to make it favour short paths. We can express this as another term to be added to the first three, of the form

$$1/2D\sum_{X}\sum_{Y\neq X}\sum_{i}d_{XY}V_{Xi}(V_{Y,i+1}+V_{Y,i-1})$$

where d_{XY} represents the distance between the two cities X and Y. The part after the d_{XY} term is non-zero only if the cities X and Y occur next to each other on the tour route, and in this case the term for the summation is equal to the distance between those cities. Thus the whole summation is equal to the length of the path for that tour.

When added together, and with the constants A, B and C sufficiently large, the really low energy states of the energy function will have the form of a valid tour. The energy of the state represents the length of the tour—the very lowest energy state will represent the optimal, shortest tour. Since the energy function contains all the information needed to solve the problem, we must provide inputs to the system that are not biased towards any one tour, so we use small random values and let the net calculate the optimal result. The inputs are not unimportant; a different starting state may well lead to a different tour, but both will be good solutions to the problem.

The results obtained from the Hopfield net are shown in figure 6.9. The most difficult problem is finding a suitable set of constants that guarantee a valid tour and allow the network to converge within a reasonably short time.

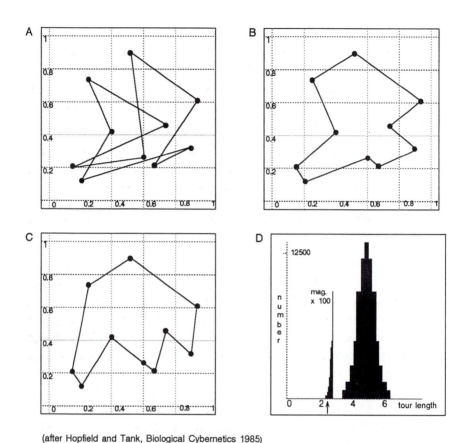

(after Hopfield and Tank, Biological Cybernetics 1985)

Figure 6.9 Results obtained in a typical travelling salesman problem, for 10 cities. 'A' shows a random route, whilst 'B' and 'C' are results obtained by the network. 'B' is also the optimal route. The histogram in 'D' shows the number of walks of a particular length that exist: the values below 3.0 have been magnified 100 times for clarity. The arrow below the horizontal axis shows the results obtained by the network.

6.5.2   The Elastic Net

There have been other approaches to optimisation tasks as exemplified by the travelling salesman problem. One of the more successful approaches is the *elastic net* of Durbin and Willshaw. The elastic net can be thought of as a number of beads connected by elastic into a ring. For the travelling salesman problem, the ring is expanded so that it satisfies two constraints:

1. Each city should eventually have a bead at its location, thus ensuring the route passes through all the cities. This is achieved by having cities pull nearby beads towards them.

2. The elastic should be as short as possible, thus minimising the distance travelled.

The cities therefore pull the beads towards them, with a force that falls off with distance like a Gaussian function, and beads pull neighbouring beads towards them. As time passes, constraint 1 is made more important than constraint 2. This is achieved by making the variance, or spread, of the Gaussian function smaller, thus pulling the bead closer and closer to the city's location. This is shown in figure 6.10.

These constraints can be expressed as terms in an energy function. The first is dependent on the distance from the city to a bead, and is the argument to a function that decreases with increasing distance. It is this function that is chosen to be the gaussian. The second constraint is satisfied by making nearby beads as close together as possible. These can be combined to give an expression for the change in position of a bead, written as Δy_j, as

$$\Delta y_j = \alpha \sum_i w_{ij}(x_i - y_j) + \beta k(y_{j+1} - 2y_j + y_{j-1})$$

where the x_i represent the positions of the cities. The w_{ij} term decreases with distance, and the second term represents the elastic tension in the net that pulls neighbouring beads together. The solutions obtained by the elastic net are generally better than those obtained with the Hopfield net, and it has the advantage that it

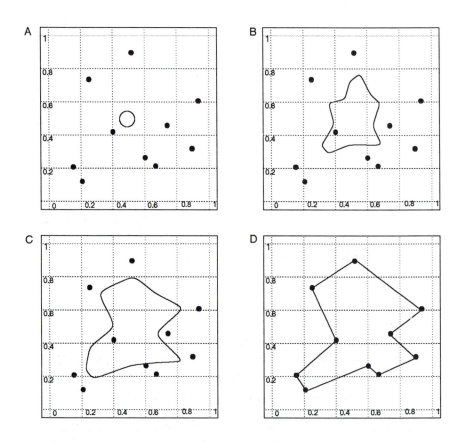

Figure 6.10 The elastic net shown as it evolves towards a solution to the travelling salesman problem

scales up to very large numbers of cities relatively well, unlike the Hopfield net.

 ## Summary

- Hopfield net—symmetric, fully connected.
- Iterates to solution.
- Acts as autoassociative memory.
- Boltzmann machine—Hopfield net with probabilistic update rule.
- Uses simulated annealing (high temperature falling to low) to assist convergence to global minima.
- Can solve constraint satisfaction problems.

FURTHER READING

1. Neural Networks and Physical Systems with Emergent Collective Computational Abilities. J. J. Hopfield. In *Proc. Natl. Acad. Sci. USA.*, volume 81, pages 3088–3092, 1984. Hopfield's original paper containing the Hopfield network.

2. A Learning Algorithm for Boltzmann Machines. G. E. Hinton, T. J. Sejnowski, & D. H. Ackley. Technical Report CMU-CS-84-119, May 1984.

7

Adaptive Resonance Theory

7.1 INTRODUCTION

Whatever their merits or failings there is little doubt that neural networks remain a controversial area within the world of computer science. Within the body of research itself, however, nobody's work is viewed more controversially than that of Dr. Stephen Grossberg. Over the last twenty years Grossberg has contributed a vast range of theory to the field covering most areas of human psychology and neurobiology. With a background in mathematics and neurobiology his work is characterised by rigorous attention to mathematical detail and accuracy. His long term research goal is to develop a unified body of theory and mathematical methodology to bring together the many diverse areas encompassed within the study of neurobiological systems. His belief is that progress in the area will be hindered until a solid underlying body of mathematics has been evolved to describe the complex dynamics of neurobiological systems.

Grossberg's work has not been restricted to analysing the dynamics of individual neural cells, but has been directed towards finding solutions to many of the neurobiological "mysteries", at what might be described as the "systems" level. More specifically, he has addressed the question of how complex systems can be developed using locally interactive and highly interconnected regions of cells. A quick scan through his published material will show that being controversial has certainly not hindered his progress—he has papers proposing models for such diverse concepts as cognition, motor control (limb movement), vision, perception and self-organisation. They are complex models, usually fully described by non-linear differential equations, based on behavioural data, and in most cases they are able to replicate many of the subtle dynamics displayed in natural biological systems.

One of Grossberg's major concerns was establishing stability in a self-organising system. Such a complex network as the brain, with its massive interconnectivity and "modular" architecture, must have a means of maintaining stability at all levels. The network we are about to discuss was developed from studies into stable neural architectures. Most neural network paradigms are plagued by a problem known as the *stability-plasticity dilemma*. This is a rather grand definition for the basic problem that networks have of not being able to learn new information on top of old. In a multilayer perceptron network, for example, trying to add a new training vector to an already trained network may have the catastrophic side-effect of destroying all the previous learning by interfering with the weight values. With training times for large networks requiring considerable amounts of computer time (hours, even days) this is a serious limitation.

Grossberg's best known work in the neural computing world is his adaptive resonance theory. It is a self-organising network that has been able to solve the stability-plasticity dilemma. This partially explains the network's high profile, although Grossberg's application of the model to pattern recognition problems has also raised interest.

7.2   ADAPTIVE RESONANCE THEORY—ART

The adaptive resonance theory (hereafter referred to as ART) was developed to model a massively parallel architecture for a self-organising neural pattern recognition network, based on biological and behavioural data. The major feature of ART, proposed by Grossberg and Gail Carpenter, is the ability to switch modes between plastic (the learning state where the internal parameters of the network can be modified) and stable (a fixed classification set), without detriment to any previous learning. The network also displays many behavioural type properties, such as sensitivity to context, that enables the network to discriminate irrelevant information or information that is repeatedly shown to the network.

The following discussion of the ART paradigm will be fairly "mechanical", in that we have curtailed much of the descriptive detail relating to psychological or cognitive effects. We have included whatever we feel adds to a fuller understanding of the workings of the network, but we have, by and large, reduced the description to a fairly basic level. We must also mention that the ART network is implemented in three versions (ART-1, ART-2, ART-3) and the following discussion will only cover ART-1 in depth.

7.3   ARCHITECTURE AND OPERATION

The ART network relies on details of architecture far more than most other neural network paradigms. The layers of the network have different functions—unlike the fairly homogeneous layers of the multilayer perceptron or Kohonen networks—and there are external parts to the layers that control the data flow through the network. Because of this, it is probably worth explaining the way that the architecture is implemented, before going on to describe the operation of the network during learning and classification.

7.3.1   The ART architecture

The ART network is shown schematically in figure 7.1.

It has two layers; the first is the input/comparison layer and the second is the output/recognition layer. We shall use the terms comparison for input and recognition for output interchangeably throughout the discussion (with apparently reckless abandon) because the functionality of the layers changes during the various cycles. These layers are connected together, again unlike most other networks, with extensive use of feedback—from the output layer to the input layer and also between the nodes of the output layer as lateral inhibition. Some of these weights are shown in figure 7.2.

This means that the ART network has *feedforward* weight vectors from the input layer to the output layer and *feedback* weight vectors from the output to the input layer. We will designate these feedforward and feedback paths W and T respectively, to avoid confusion.

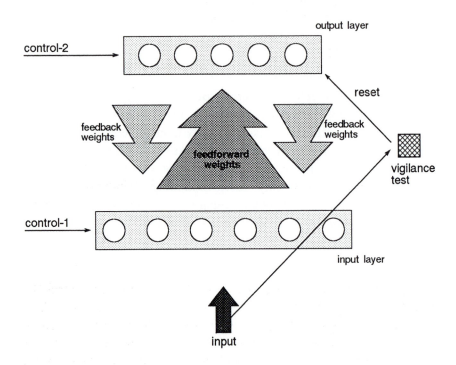

Figure 7.1 The ART architecture.

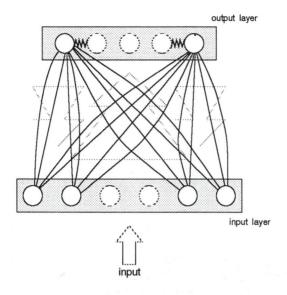

Figure 7.2 The weights within the ART architecture.

For each layer there are also logic control signals that control the data flow through the layers at each stage of the operating cycle—these we will designate control-1 and control-2. The respective inputs are common to each node in the input and output layers. Between the input and output layers there is also a reset circuit. This plays a vital role in the network; it performs more than simply a reset function for the output nodes—it is actually responsible for comparing the inputs to a vigilance threshold that determines whether a new class pattern should be created for an input pattern.

This is the basic architecture of the ART network; the points to note are the extensive feedback connections, the separate functions of each layer, and the external control signals—all of these will now be explained in operation.

7.3.2   ART-1 Operation

There are several phases to learning or classification in an ART network. The most obvious difference from most other network paradigms is that the continually modified input vector is passed forwards and *backwards* (resonated) between the layers in a cyclic process. We shall describe the action of the network in terms of the activity at the separate layers for each phase. These phases can be broadly divided into an initialisation phase, a recognition phase, a comparison phase and a search phase.

7.3.3   The Initialisation Phase

The initialisation of an ART network requires more work than is the case for most other neural networks, which is perhaps not surprising because it has more features to it than most others. The two control signals, control-1 and control-2, direct the data flow through the network during the various learning or classification phases. Control-1 determines the course of data flow for the input layer—its binary value toggles the first layer of nodes between its two modes; input and comparison. The state of control-1 is *one* whenever a valid input (i.e. non-zero) is presented to the network but is forced to *zero* if any node in the recognition layer is active. Control-2 is the simpler of the two—its binary value enables or disables the nodes in the recognition layer. It is *one* for any valid input pattern but *zero* after a *failed* vigilance test (this disables the recognition layer nodes and resets their activation levels to zero).

The weight vectors, W and T, must also be initialised. The feedback links are simple; they are all set to binary one, inferring that every node in the output is initially connected to every node in the input via a feedback link. The feedforward links are set to a constant real value determined by:

$$w_i = \frac{1}{1+n}$$

where n is the number of input nodes.

The vigilance threshold is also set in the range $0 < \rho < 1$. The significance of this will become apparent during the discussion of the operating cycle.

7.3.4   The Recognition Phase

In the recognition phase the input vector is passed through the network from the input layer and its value is matched against the classifications represented at each node in the output layer. We shall discuss how the recognition layer nodes adopt these classifications during the training cycle.

The nodes in the input layer each have three inputs: a component of the input vector x_i, the feedback signal from the output layer, and the control-1 signal. Data flow through the input layer is controlled by the "two-thirds" rule suggested by Grossberg and Carpenter—if any two inputs to a node are active then a one is output from the node, otherwise the node is held at zero output.

The recognition phase has parallels with the Kohonen network discussed in Chapter 5. Each weight vector W at each recognition node can be thought of as a "stored template", or exemplar class pattern. The input vector is compared to the exemplar at each node and the best-match is found. The best-match comparison is done by computing the dot product of the input vector and a node's weight vector—the node with the closest weight vector to the input will yield the largest result. Several nodes in the recognition layer may in fact respond with a high level of activation due to the input vector, but the lateral inhibition between the nodes now comes into play, turning off each node except the maximum response node. This node will inflict the largest inhibitory effect on the other nodes, so that although all the nodes are actually trying to turn each other off, it will be the maximum response node that dominates the effect. Each node also has positive feedback to itself to reinforce its own output value. The combined effects of reinforcement and lateral inhibition will ensure that only one node remains significantly active in the layer.

The winning node is now required to pass its stored class pattern (T—the class exemplar) back to the comparison layer. If we recall that the exemplar is actually stored as a binary weight vector in the feedback links to the input layer, we can see that the exemplar can actually be passed to the comparison layer by simply mapping the winning node's activation (which is forced to one by the action of the positive reinforcement) through the feedback weights to the input layer. If this is difficult to visualise then consider the diagram of figure 7.3 that shows just the feedback links from the winning node in the recognition layer to the input layer.

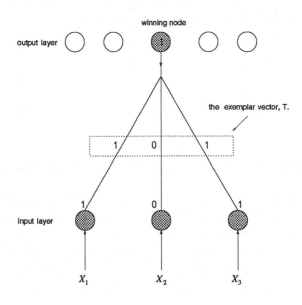

Figure 7.3 It is the feedback vector that stores the exemplar vector.

7.3.5   The Comparison Phase

Two vectors are present at the input layer for the comparison phase—remember that each node in the input layer has three inputs. On one input of each node the input vector is clamped and on the second input the exemplar vector from the recognition layer is clamped. The third input is the control-1 signal which is zero for the duration of this phase because the recognition layer has a fully active node. The situation is depicted in figure 7.4.

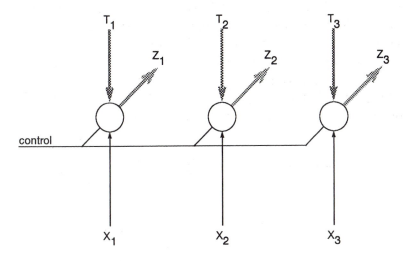

Figure 7.4 The input layer with all three inputs applied.

The *two-thirds* rule applies for calculating the output of each node. The exemplar vector and the input vector are thus ANDed together (control-1 is zero and so has no effect on the output) to produce a new vector on the output of the comparison layer. This we will call the *comparison vector* and designate Z. The comparison vector is passed to the *reset* circuit along with the current input vector.

7.3.6 Vigilance Threshold

The reset circuit is responsible for testing the similarity of the input vector and the comparison vector against the vigilance threshold. The test is a ratio count of the number of ones in both the input vector and the comparison vector. It is not a difficult ratio to evaluate—the dot product of the comparison vector and the input vector will yield a count of the matching ones in each pattern. This is divided by the bit count of the one bits in the input vector to provide a ratio, S, which is subsequently compared to the vigilance value.

$$S = \frac{\sum t_{ij} x_i}{\sum x_i}$$

test: Is $S > \rho$

If S is greater than the vigilance threshold, ρ, then the classification is complete and the class membership is indicated by the active node in the output layer. If the ratio is below the threshold then this implies that we have not found the correct best-match exemplar, and the network enters the search phase.

7.3.7   The Search Phase

During the search phase the network is attempting to find a new matching vector in the recognition layer for the current input vector. First the present active output node is disabled and its output zeroed. This has a twofold effect: the node is prevented from entering any further best-match comparisons for the current input pattern, and the control-1 signal is forced to zero, since the outputs of the recognition layer are again all zero. The input vector is now reapplied to the recognition layer and the best-match comparison is recalculated as described above. The network enters the comparison phase again, which ends with the new recognition layer exemplar being tested against the vigilance threshold. This process is repeated, consecutively disabling nodes in the output layer, until a node is found in the recognition layer that matches the input to within the limits of the vigilance threshold. If no such node is found then the network makes the decision to declare the input vector an unknown class and allocate it to a previously unassigned node in the output layer.

This completes the working description of the various stages of the ART network and explains how data is dynamically routed around the network in a "resonant" fashion. The term resonant is most appropriate, because of the way in which the input vector is "bounced" back and forth between the input and output layers before it finds a stable state. As you can see, there is a good deal more complexity in the ART network than in the majority of other neural network algorithms. The algorithm itself, however, is neither notionally difficult nor computationally complex. It can be implemented, as suggested by Lippmann, in the following fashion.

7.4   ART ALGORITHM

 The ART algorithm is given below.

The ART1 Algorithm

---

1. Initialise

$$t_{ij}(0) = 1$$
$$w_{ij}(0) = \frac{1}{1+N}$$
$$0 \le i \le N-1 \qquad 0 \le j \le M-1$$

Set ρ, where $0 \le \rho \le 1$
where $t_{ij}(t)$ is the top-down and $w_{ij}(t)$ is the bottom-up connection weight between node i and node j at time t. It is these weights that define the exemplar specified by output node j. ρ is the vigilance threshold which determines how close an input has to be to correctly match a stored exemplar. There are M output nodes and N input nodes.
2. Apply new input
3. Compute matching

$$\mu_j = \sum_{i=0}^{N-1} w_{ij}(t)x_i$$
$$0 \le j \le M-1$$

μ_j is the output of node j and x_i is element i of the input which can be either 0 or 1.

4. Select best matching exemplar

$$\mu_{j*} = \max_j[\mu_j]$$

5. Test

$$\|X\| = \sum_{i=0}^{N-1} x_i$$

$$\|T \cdot X\| = \sum_{i=0}^{N-1} t_{ij*}(t)x_i$$

is $\dfrac{\|T \cdot X\|}{\|X\|} > \rho$

YES   go to 7

NO   go to 6

6. Disable best match

Set output of best match node to 0. Go to 3.

7. Adapt best match

$$t_{ij*}(t+1) = t_{ij*}(t)x_i$$

$$w_{ij*}(t+1) = \frac{t_{ij*}(t)x_i}{0.5 + \sum_{i=0}^{N-1} t_{ij*}(t)x_i}$$

8. Repeat

Enable any disabled nodes, then go to 2.

7.5 TRAINING THE ART NETWORK

The training cycle for the ART network has a different learning philosophy to other neural network paradigms. The learning algorithm is optimised to enable the network to re-enter the training mode at any time, to incorporate new training data. As we discussed earlier, this is a practical solution to the stability-plasticity problem, and the ART network is possibly one of the only neural networks that can cope with learning in a continually varying environment. The following discussion will describe the factors that affect the learning performance of the network.

There are, in fact, two training schemes for ART, which are described as *fast learning* and *slow learning*. Fast learning is so called because the weights in the feedforward path are set to their optimum values in very few learning cycles—in fact, in most implementations, they are learnt in a single pass of the training data. Conversely, slow learning forces the weights to adapt slowly over many training cycles. The advantage of this technique is that the weights are trained to represent the statistical *average* of the input data for any particular class. This means that more attention will be given to finding the salient features of the input patterns that determine the classifications. Generally it seems that fast learning is the method most often adopted—although this may be for no other reason than that it is simpler to implement.

ART is very sensitive to variations in its network parameters during the training cycle. Undoubtedly the most critical parameter is the vigilance threshold, which can dramatically alter the performance of the network. Also important is the initialisation of the feedforward weight vectors—they must all be set to low values at the start of training. If any vector is not initialised to a small value it will dominate the training process, because it will invariably win the best-match comparison at the recognition phase. This means that all the input vectors will be assigned to just one output node—by any stretch of the imagination that is a broad categorisation process! Consequently, the algorithm forces all the weights to small, equal values during initialisation.

The vigilance parameter controls the resolution of the classification process. A low choice of threshold (< 0.4) will produce a low resolution classification process, creating fewer class types. Conversely, a high vigilance threshold (tending to 1) will produce a very fine resolution classification, meaning that even slight variations between input patterns will force a new class exemplar to be made. In many cases, a high value will make the network too sensitive to dissimilarities between inputs of the same class, and will quickly assign all the available output nodes to new classes. A major criticism of the ART network is that it performs poorly in noisy input conditions because of this vigilance problem. However, it must be noted that this is not an oversight on Grossberg's part, rather it is an attempt to make the network's performance sensitive to its environment. By this we mean that context is taken into account; depending upon the circumstances, a discrimination problem can demand coarse or fine categorisation. We know from our own experience that in some circumstances we are quite willing to accept a very broad generalisation of a concept, whereas in other circumstances it would be highly undesirable to have anything less than a very accurate delineation. As an example, consider the vast multitude of shapes and sizes that we include under the category of *table*. We don't learn the specific features of every table we see, in order to be able to recognise a table. Conversely, learning telephone numbers would be of little use to us if we simply remembered that they need six digits and (sometimes) a code number.

Ultimately, whether the network's sensitivity to the vigilance parameter is an advantage or a failing depends upon the perspective in which the role of ART is seen. As a model of contextual sensitivity to training data, it performs in a very plausible manner when compared to behavioural data; as an engineering tool to perform pattern recognition, it has severe drawbacks.

To illustrate the sensitivity of the network to changes in the vigilance parameter we include the following examples modelled after Grossberg's experimental results, figure 7.5.

With a low vigilance threshold ($\rho = 0.2$) the number of bits in the vertical stroke of the input pattern is enough to warrant each of the input vectors to be assigned to the same node at the output layer and consequently the same class. With a high vigilance threshold ($\rho = 0.8$), the features of the input patterns are examined much more carefully and they are considered to be sufficiently different for a unique class to be assigned to each.

The optimal solution may be to vary the vigilance value dynamically during the training process. A low initial value would quickly

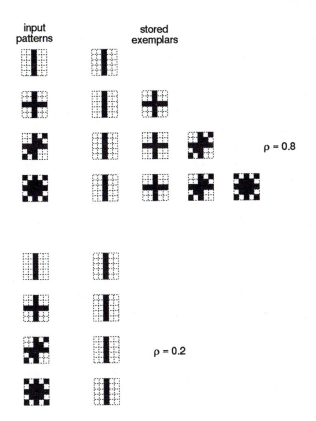

Figure 7.5 The classification performance of the network is controlled by the vigilance parameter, ρ. Two examples of training, with different values of ρ, are shown.

assign the coarse clustering of the input patterns and increasing this later in the training cycle may optimise the classification. In keeping with his drive to model real cognitive processes, Grossberg describes the possibility of modifying the vigilance value during training as a "punishment event". By that Grossberg means that if the network makes an erroneous classification, then the network should be "punished" for it. Punishment takes the form of negative reinforcement which amplifies the activity passed to the reset circuit and subsequently modifies the value of the vigilance parameter. The punishment must be administered by an external circuit that monitors the response of the network; this implies that the network is now using *reinforced* learning rather than unsupervised.

7.5.1   Scaling the Feedforward Weights

The ART model includes a process in the learning algorithm that incorporates what Grossberg describes as "self-scaling" of the feedforward real valued weight vector, W. The effect of this process is as critical to the classification performance of the network as the vigilance parameter, since it makes a step towards distinguishing noise from the signal in an input vector. We can explain the scaling process by looking at the equation for adapting the weights in the feedforward path:

$$W_{ij} = \frac{LZ_i}{(L - 1 + \sum Z_k)}$$

The term Z_k in the denominator is equal to the number of active bits in the comparison vector (because the vector is binary). Consequently, we can see that all the weight components, Z_{ij}, are "normalised" by the active bit count of ones in the comparison vector. This causes the comparison vectors with a high number of bits set to one to produce smaller weight values than those with a comparatively low active bit count. The effect this has on the classification process is best explained by the following example. Consider two input patterns, representing different classes:

$$x_1 = 1\ 0\ 0\ 0\ 0\ 0$$
$$x_2 = 1\ 0\ 0\ 1\ 1\ 1$$

If scaling was not used during the learning stage then the feedforward weights would be set to the same values as the feedback weights, where no scaling is used:

$$w_1 = x_1 \; = \; 1\;0\;0\;0\;0\;0$$
$$w_2 = x_2 \; = \; 1\;0\;0\;1\;1\;1$$

If x_1 is applied to the network again after training, the response of nodes one and two in the recognition layer will in fact be the same (the dot product of x_1 with w_1 or w_2 is the same in either case). Either node is, therefore, equally likely to win the best-match comparison. If node 2 wins then the network is in trouble! Apart from the input being erroneously classified, the exemplar for node two will in fact be corrupted since it will be modified to follow the form of the input vector x_1—thus undoing its previous training.

However, using scaling during training will in fact create the following feedforward weight values:

Let L=2;

$$W_1 \; = \; 1 \quad 0 \quad 0 \quad 0 \quad 0 \quad 0$$
$$W_2 \; = \; .4 \quad 0 \quad 0 \quad .4 \quad .4 \quad .4$$

Reapplying input vector x_1 will now produce a different dot product value for each weight vector and, in this case, x_1 will only activate node one, to produce the correct classification.

Summing up these results we can see that scaling prevents any vector that is a subset of another being classified in the same category. The consequence of this is that two vectors that share common features, but are in different classes, can still be distinguished. Grossberg describes the action of the self-scaling technique as the discovery of critical features in a context sensitive manner.

7.5.2   The Training Cycle

For completeness we will work through a training cycle, describing how the input vector is passed through the stages of the network before finally being assigned to an output node. We will assume that the network has three input nodes, an arbitrary number of,

Table 7.1 The two-thirds rule.

Input	Recognition layer	Control-1	Result by 2/3 rule
1	1	1	1
1	1	0	1
0	1	0	0

say, twenty output nodes and is initialised: control-1 and control-2 are both zero, the output layer is all zero and the weight vectors are in their starting states; feedforward weights are set to a value determined by:

$$w_{ij} = \frac{1}{(1+N)} \quad \text{where } N = \text{dimensionality of the input vector.}$$

and the exemplar patterns, stored in the feedback weights, are all set to binary one. The input vector, X_1, can now be applied to the input layer. The "two-thirds" rule determines the response of the layer to the input pattern; at this point we have only two active inputs on the input layer; the input signal and the control-1 signal which is binary one (signifying that there is a valid input to the network). This has the effect of ANDing the input vector with the control-1 signal which means that the input vector is passed unchanged through to the next layer.

This layer is of course the recognition layer, where the input vector is matched against the feedforward vectors at each node by calculating the dot product of the input and weight vectors. However, because all the feedforward weights are initialised to the same starting value, it will be an arbitrary choice as to which is selected as the best-match. The node selected as the winner in the recognition layer passes its stored exemplar back to the input layer, and the control-1 signal is forced back to zero. The input layer now has three inputs—the input vector, the exemplar vector and the control-1 signal. The output of the layer by the two thirds rule is shown in table 7.1.

The comparison vector $(1, 1, 0)$ and the input vector $(1, 1, 0)$ are now both passed to the reset circuit for the vigilance test. The similarity ratio for the two vectors is evaluated—which in this case is quite simply 1 : 1 because the vectors are identical—and the result is compared to the vigilance threshold. The vigilance is 0.8 so the similarity ratio is above the threshold value and the input vector is assumed to be correctly classified. Once the vigilance test is passed, the winning node weight vector is updated to incorporate the features of the input vector. This is done by ANDing the old exemplar vector with the current input:

$$T_{ij} \text{ new} = T_{ij} \text{ old} \wedge X_i$$

where $\wedge$ is the logical AND operator.

Thus, for our input, the winning node will have its exemplar modified to:

$$
\begin{aligned}
T_{ij} \text{ new} &= (1,1,1) \wedge (1,1,0) \\
&= (1,1,0)
\end{aligned}
$$

The input vector, X_1, is now stored as a class type at the node in the recognition layer.

If we now apply another training input to the network, $X_2 = (1,0,1)$, and recalculate the matching scores at the recognition layer, we will find that the node assigned to the X_1 input will be the winning node. This is because its feedforward weight values are much larger than those of the other, as yet, unassigned nodes. As a result the exemplar for class 1 (input X_1), will be passed, erroneously, to the comparison layer with the input X_2. However, if we trace the exemplar through to the reset circuit as before, we obtain the following result:

$$S = \frac{\sum T_{ij} X_i}{\sum X_i}$$

$$S = 1/2 = 0.5$$

Now $S < \rho$ so the network decides that, although node one was chosen as the best-match, it is actually a wrong classification and the network enters the reset phase. This means that node one will be disabled (for the duration of the present input), the recognition layer reset to all zero and the vector reapplied to the recognition phase without node one. In this case the classification will proceed as for the first input and X_2 will be assigned to a unused output node.

One important feature to note is that the learning time for the network is much faster than the iterative convergence procedures proposed for most other neural networks. When learning a new pattern the slowest part of the process is actually performing the search in the recognition layer. However, even this process is not slow in comparison to other neural learning paradigms because the search process is actually performed in parallel. The best-match comparison is computed simultaneously for each node in the recognition layer rather than sequentially. The most important feature is that none of the weight values are modified at all until the search process has halted and one node has been selected. Using *fast learning* the weight values are modified to update the classification to a perfect match in just one presentation of the input. Any subsequent learning will refine these classifications (still in one pass) by incorporating more features found in the class training examples. We must once again stress, however, that the performance issues are still heavily dictated by the choice of vigilance threshold, which has almost total control of the network's generalisation and classification properties.

We made the point earlier in the chapter that it is somewhat ambiguous to call this process a learning *cycle* since the learning mechanisms that we have described stay intact throughout the operation of the network. This implies that whenever a new input is presented to the network during the classification process if no suitable matching classification is found then one is added to the recognition layer. The only limit to this process is the number of nodes that remain uncommitted in the layer—the search and learning process will always terminate on an unassigned node. If no nodes are available then the input will remain unclassified. In practice, the learning process does in fact settle to a steady state, as a significant number of classifications are formed at the recognition layer, since the likelihood of a new input matching one of the known classes will increase.

7.6   CLASSIFICATION

The ART network exploits to the full one of the inherent advantages of neural computing techniques; namely parallel processing. It models the mechanisms that allow the human brain to perform recognition rapidly despite the apparently prohibitive size of the knowledge base that has to be searched. Furthermore, despite the vast number of internal representations in the brain, encoding abstract knowledge concepts, there is no evident conflict in the recognition or recall of familiar objects. This would appear to indicate that there is little plausible evidence for the brain using such methods as semantic nets or sequential tree structures to represent data internally. This conjecture—controversial as it may be in some quarters—is upheld by Grossberg and others, and stresses the need for parallel search methods. In this respect Grossberg and Carpenter make two claims for the performance of the ART network. The first of these is that, despite the size and complexity of the encodings in the recognition layer, familiar input patterns (which implies those classes of input used to train the network) will have direct access to the classification nodes in the output layer. The second claim is that the network uses a self-adjusting memory search that will optimally search the recognition layer, in parallel, to classify an unfamiliar input. We have broached these points already in our discussion of the search and recognition phases. The classification of any input is done in an inherently parallel fashion since the input vector is presented to each of the nodes in the recognition layer simultaneously. This has the obvious implication that the technique can be made parallel at the implementation stage; however, this was not the primary concern of Grossberg. He has attempted to show how mechanisms to allow a parallel search may be implemented at the neuron level. Similarly for the idea of direct access to recognition nodes. Any unfamiliar input pattern will still activate a node in the recognition layer if it shares enough salient features to patterns learned previously. As we have already described, how close the features have to be is determined by the level of the vigilance threshold (which can be likened to a control parameter that moderates sensitivity to context in the training data).

The most important point to note about the network in the classification stage is that it remains open to adaption in the event of new information being applied to the network. If an unknown input is applied to the network ART will always attempt to assign a new class in the recognition layer by assigning the unknown input to a node. The prohibitive limit to this process is the number of nodes available in the recognition layer. As we discussed earlier it is the ability of the network to switch between stable and plastic states without detriment to the performance of previously learned data or to the speed of classification that makes it a unique example of a neural network. It can again be considered a natural embodiment of the human learning process which cannot be described as having a learning cycle (unless we make it three score and ten!) and is perfectly adapted to merging new experience with old.

7.7   CONCLUSION

That completes an operational overview of the ART network. Through it we have attempted to remove some of the "mystique" and confusion surrounding the implementation of adaptive resonance theory, by providing a "nuts and bolts" description of how the network operates. We did mention earlier, that there are three models of adaptive resonance theory, called, not surprisingly, ART-1, ART-2 and ART-3, and this discussion has only covered ART-1. ART-1 and ART-2 are actually very similar, the major difference being that ART-2 is a "real valued" implementation of ART-1. By that, we mean the input layer takes real valued vectors as inputs, as opposed to binary vectors in ART-1. A typical example of a real valued input might be a grey scale pattern obtained from an image processing system, where the elements of the vector are usually discrete values in the range 0–255.

The architecture for ART-1 and ART-2 are basically the same, but there are subtle differences in the implementation of the input layer to deal with the use of real valued vectors. The input layer has also been split into several functional layers so that much more complex matching of recognition layer and comparison layer data can be achieved. This incorporates such effects as feature enhancement, noise suppression, sparse coding and expectation from the recognition layer. Positive feedback is also used between the buffers of the input layer. The performance of the ART network is vastly improved and it has been applied to applications such as pattern recognition, speech perception and radar classification.

ART-3 uses the same network topology as ART-2, but it uses equations that model the dynamics of chemical neurotransmitters. Grossberg and Carpenter have turned their attention to mapping the functionality of the ART model onto a representation of a biological neural architecture. In doing so they have also countered a major criticism of ART-1 and ART-2: that the network did not use a distributed representation for the internal coding of the categories. It also means that the input and output layers of the network are similar because they use the same node model. The significance of this is that the network can now be modularised, such that the output layer of one network feeds directly into the input layer of another, enabling hierarchies of networks to be built. ART-3 also accepts real-time constantly varying inputs, the input is continually monitored and when the signal changes significantly a reset phase is triggered that searches the recognition/learning cycle. This is probably the closest a network has come to modelling both the architecture and the dynamics of a biological neural network.

7.7.1   Terminology

It is worth mentioning before we close this chapter that we have not used Grossberg's terminology during our discussion of ART. The main reason for this is that Grossberg's description of parts of the network are couched in "psychological" phraseology, and we thought it would be of more benefit to avoid this extra confusion. We shall endeavour to put the record straight now though, and explain Grossberg's terminology, primarily for the benefit of those who wish to read further into his work, and still relate back to the description we have given here.

The most significant difference is Grossberg's definition of the weight vectors. For simplicity, we have labelled them as the feed-forward and feedback weight vectors, which we hope is fairly self-explanatory. Grossberg, however, prefers to describe the weights as *memory traces*. The stored exemplar vector, T, and the feed-forward weights W, he describes as *long term memory* traces—the analogy is quite clearly drawn from his interest in biological systems. The exemplar vector is "locked" into memory as a consequence of learning—barring minor updates to this data as new information comes along, we require this information to be stored long term and in a stable state. The short term memory traces correspond to transient states of the network, in other words, the activity at the recognition and comparison layers. These states are not stored, they are continually modified during the learning process as the memory is searched for matching information. Once a stable output state has been found, these short term memory traces are reset, ready for new information to be presented. One other minor point about the weights is that Grossberg describes the feedforward connections as a bottom-up *adaptive filter*. This is simply another way of thinking about the transform of the input vector through the weight matrix, and because the ART model is based on cognitive effects, it is per-haps more useful, in some circumstances, to think about the weights "filtering" the information that is passed through them.

The comparison layer and the recognition layer are described as a *feature representation field* , $F1$, and a *category representation field* , $F2$, respectively. These are intuitive labels that describe the functionality of the layers during the learning/classification cycle.

The control signals, that we have called control-1 and control-2, are labelled as *attentional gain control channels* by Grossberg. The reason for this is that Grossberg chooses to describe the function of these signals in terms of subtle cognitive effects, such as *subliminal activity* and *attention priming*. These effects are modelled on cogni-tive or behavioural data, and although interesting in themselves, we did not feel that they were within the scope of this text.

7.8   SUMMARY OF ART

The ART network has many significant differences from other neural paradigms. The most notable achievement of the ART model is the ability to deal with the stability-plasticity dilemma of learning in a changing environment. The network will continue to add new information, until it utilises all of the available memory, and will continually refine the knowledge stored within it as new information is presented. The network has been rigorously proven to be stable and does not suffer from any convergence problems such as local minima. The learning algorithm is unsupervised and requires only one pass through the training set to learn the internal representations (if fast-learning is used). ART can also deal with both binary or real valued inputs under the ART-1 or ART-2 guises.

The criticisms of ART (ART-1—later models have significantly improved the performance and plausibilty of the network) are aimed at the poor results in noisy input conditions, the use of non-distributed coding of data (i.e. ART uses the "Grandmother" cell approach), and the implausible "neural" architecture of the network—despite it being based on biological studies.

 Summary

- ART is an unsupervised, vector-clustering, competitive learning algorithm.
- ART has provided a solution to the stability-plasticity learning dilemma.
- ART is fully described mathematically by non-linear differential equations.
- ART is based on cognitive and behavioural models.
- ART uses extensive feedback between input and output layers.

- ART is implemented for both real and binary inputs.

Further Reading

1. *Neural Networks and Natural Intelligence.* S. Grossberg. MIT Bradford Press, 1988. The definitive collection of papers from Grossberg's group.

2. The ART of Adaptive Pattern Recognition. G. A. Carpenter & S. Grossberg. In *IEEE Computer*, volume 21, number 3, March 1988. An introductory paper to ART. Useful discussion and a good source of further references.

8

Associative Memory

Associations are common within our everyday experience. We are easily able to put names to faces, to recall that someone looks familiar because they work with us, and so on. We form links between people, events and places, between shapes and objects and concepts, and this ability allows us to build our own representation of the world as we see it. Inputs to our senses usually trigger off a cascade of associations and recollections, each one prompting the next; a piece of music may evoke memories of warm summer evenings and images of a particular person, or a barking dog may make us smile at some childhood incident. It is clear that human memory works in an associative fashion, but, in more general terms, we can describe associative memory as a memory system such that an input specifically evokes the associated response.

Computational models of associative memory have been studied for many years, and much of the work in neural networks draws on the ideas developed in this field. The distinction between an "associative memory" and a "neural network" is imprecise, and is often a matter of personal preference since many networks operate as associative memories (for example, the Hopfield network associates patterns with themselves), whilst some associative memories perform the same processing as a network.

8.1 STANDARD COMPUTER MEMORY

Associative memory appears familiar to us since it corresponds to the way in which our own memories operate. However, the memory of a conventional computer does not work in the same way. In a standard computer memory, each piece of distinct information is stored in its own section of memory, and is accessed by knowing the value of its location, i.e. its address. This local storage of information needs some form of address decoder in order to designate or retrieve the information. It is like wanting to send a friend a letter; you may well know their name, but to get the letter to them you have to search through your address book to find their address, send the letter to that address, and then they will receive it. Unfortunately if we only know their name, but have no address, we cannot send them the letter. We can extend the analogy further—just as your friend lives in a house at one address, so another completely different person lives next door, at a different address. Likewise in a computer memory, one piece of information is stored after another, each at a different address. Each piece of information is quite likely to be unrelated to the information on either side of it, just as people in adjacent houses often have nothing in common. This type of memory, where the information is stored sequentially, is called a *listing* memory, since the information is stored as a list. When recalled, the same information is reproduced in the same sequential fashion as you pass down the list. A simple example of a listing memory is a tape recorder.

Associative memory, however, requires us to associate some response to a particular input, so that when we present that input, we get the required output. It would be possible to produce a long list that contained all the inputs and their corresponding outputs, and then scan it looking for the correct input match and so the corresponding output, but this seems excessively complicated. Not only do we have to record all the questions as well as the associated answers, but we also have to move down the list each time. Instead, if we consider both the input and the output to be a patterned signal, we can envisage associating the two patterns by transforming the first into the second. The memory would only have to hold the required transformations, and not an explicit list of input–output pairs. In other words, we consider the input and output as vectors, and associate them by producing a matrix that transforms the input vector into the output vector. This matrix holds a mapping from one specific stimulus onto the associated response, and so is known as a *mapping* memory. The important point is that the input and response can each be represented by a patterned signal, and the mapping transforms one pattern into the other; there is no direct correspondence between individual elements of the patterns, only between the patterns as a whole.

In associative memories, the aim is for the presentation of one set of input signals to elicit the recall of another set of signals from the memory. This implies that the input signal contains all the necessary information to access the stored pattern, without the need for any decoding. The idea of accessing the memory on the basis of the structure of the input pattern gives rise to the term *content-addressable memory* (CAM).

With standard memory access, knowing part of address of the object to be retrieved is useless. Content-addressable memory, on the other hand, is able to recall the complete description of an object despite only having part of the input available. This tolerance to input noise makes these types of memories useful for pattern completion tasks, and for closest match recognition of unseen inputs.

Since there is no direct correspondence between the input and the response, there is no one memory location that specifically defines the output; the whole of the matrix is involved. A memory with this non-localised representation is known as a *distributed* memory. In a distributed memory, each memory element holds traces of many stored items, and it is only when viewed collectively that these individual elements form a coherent whole. There are many advantages to distributed memories since the same properties that are in neural nets are applicable. Due to the non-localised storage of the mapping, no single part is of critical importance to the overall transformation, and so the matrix is resistant to damage. This is untrue in a conventional memory. Distributed memories are also tolerant of faults, either in the memory matrix itself or in the input patterns. Again, this is due to the fact that it is the overall pattern that is important and a few isolated errors are negligible. However, it is important to remember that recall is only possible if the memory can produce a selective response to the input. Since the memory contains many mappings from one pattern to another, it must be able to separate the required output from the corrupting overlap of the other patterns.

So far, we have discussed the association between the "input" and the "output" without reference to any particular type of input or output, but in fact there are two types of association depending on the nature of the two patterns to be associated. We have already seen that we can tolerate errors in the input since the overall effect will still be sufficient to allow recall. With this in mind we can associate a pattern with itself by making the input and response patterns the same, whereupon presentation of an incomplete pattern on the input will result in the recall of the complete pattern. Recall of this nature is called *autoassociative*. If the input pattern is taught in association with a different output pattern, then the presentation of this input will cause the corresponding pattern to appear on the output; such a memory is termed *heteroassociative*. This is shown in figure 8.1.

8.2   IMPLEMENTING ASSOCIATIVE MEMORY

The question is, given that associative memory appears superior to conventional memory, can we actually implement such a system on a computer?

Let us first consider the implementation of content-addressable memory, and then examine whether this is in fact associative memory. A simple form of content-addressable memory can be implemented in standard computer memory using a technique known as *hash coding*. In hash coding, the address for storage is made a function of the item to be stored, and is determined by some mapping algorithm. For example, suppose we want to store pairs of words like

- shopping list
- tea time

We can choose to hash code these pairs on the first two letters of the first word to produce the address, making it equal to the sum of the

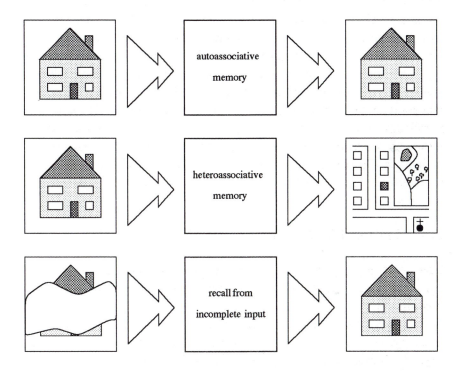

Figure 8.1 The different types of association: autoassociation and heteroassociation. The use of an autoassociative system for pattern completion is shown in the bottom figure.

alphabet positions of the letters. Given that 's' is the 19th letter in the alphabet, and 'h' the 8th, the word pair "shopping list" would be stored at location $19 + 8 = 27$, whilst t=20, e=5 means that "tea time" would be stored at location 25.

For recall, the same algorithm is applied to the input in order to recover the location. The data is scattered throughout the memory area, its position dependent only on its contents and not in any regular order. However, there are problems with hash coding. The item that is used as the hash code is known as the key; this key has to be unique since only one address can be computed from each key. Different associations with the same key would cause a clash since they would both try to occupy the same storage location, and only one item can be held at any one address. Collisions can also occur when the computed address of two different keys happens to be the same. Returning to our example, we would not be able to store the extra item

- shoe polish

since the 'sh' address will be the same as for "shopping list". Likewise,

- pink tablecloth

causes problems, since with p=16, i=9, it collides with "tea time" at address location 25.

So, does this implement true associative memory? Hash coding does provide content-addressable memory, decoding the input to provide the response, but it requires the key word to be known exactly. Our ideal associative memory should provide recall on the basis of incomplete, noisy and distorted input cues, so we need to consider whether there is a better way.

8.3   IMPLEMENTATION IN RAMS

Associative memories can be implemented in random access memories (RAMs), an approach which has been pioneered by Professor Igor Aleksander. The elements of a random access memory are shown in figure 8.2.

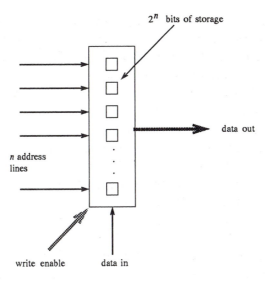

Figure 8.2 The elements of a basic random access memory.

There are n address lines, each taking a binary value of 1 or 0, so there are 2^n distinct address patterns at the input, each accessing one address. Each address can store one bit of information which would appear on the data-out line when accessed. This is known as the "read" mode, and is different to the "write" or "teach" mode. The teach mode is entered by activating the write-enable terminal, which, as its name implies, allows data to be written into the RAM. In this mode, the contents of the addressed location can be changed to the logical value determined by the data-in terminal, i.e. $+1$ or 0. This RAM can act as a simple pattern recogniser; if the pattern is applied at the n binary inputs, it can be taught by energising the write-enable input and setting the data-in line to one. These n inputs together produce a unique address, which is used to store the data-in value of $+1$. In the recognition stage, when the RAM is in the read mode, the RAM will output this 1 if the same addressing pattern occurs on its inputs. However, the RAM will only respond to those patterns on which it has been taught and will not extend the recognition to other similar patterns. Also, it requires a complete pattern on its input. This appears to be no better than using hash coding, but networks of RAMs are able to act in a more complex way, as shown by the following example.

Figure 8.3 shows an arrangement where a 3 by 3 matrix is connected to three RAMs.

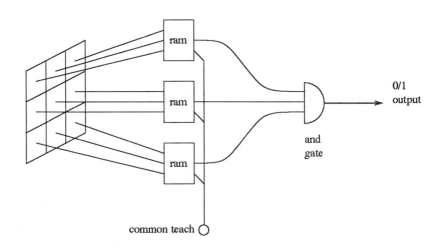

Figure 8.3 Matrix connected to three RAM units.

The training set shown in figure 8.4 is presented. The RAMs are taught to respond with a 1 for those patterns that are in the training set. Since the results of the RAMs are passed through the "and" gate, only those patterns causing all three RAMs to respond positively would be classified in the same way as the training set.

Looking at the training set, we can see that the three RAMs look at one row of the pattern each, and, during training, each RAM sees only two different addressing sub-patterns. Each ram will output a 1 when its sub-pattern occurs, so the net will recognise any possible combination of these three rows. This means that the net will recog-

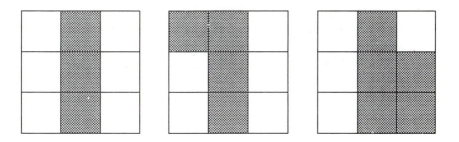

Figure 8.4 Training set for three-RAM net.

nise all of the patterns shown in figure 8.5 since these are all made up of combinations of the training set sub-patterns.

Since there are $3 \times 3 = 9$ locations in the grid, each of which can be in one of two states, either 1 or 0, there are $2^9 = 512$ different possible patterns that can be represented on the grid. Three of these were presented in training, and we get recognition of these three, plus the five additional ones shown in figure 8.5. They are similar, since they have at most one bit set differently from one of the training patterns. The net *generalises* from the taught patterns to include these other similar patterns, which are therefore collectively known as the *generalisation set.*

The ability of the recogniser to generalise is an important feature of such a system, and the size of the generalisation set is controlled by the diversity of patterns in the training set. If the number of subpatterns seen is greater, so will be the size of the generalisation set, since there are more possible combinations of the sub-patterns. The gate used to combine the RAMs output for the output decision is also crucial, since all RAMs have to see a known sub-pattern for classification to occur with an AND gate, whilst only one has to respond for classification to occur if an OR gate is used. These combinations of RAMs are known as single-layer RAM nets. Their generalisation properties are summarised in figure 8.6.

This simple architecture divides the set of all possible patterns

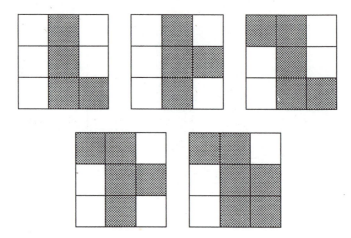

Figure 8.5 The generalisation set: extra patterns that the network recognises.

into those that are in the generalisation set, and those that are not. Most pattern recognition problems require more categories than this, and so many RAM nets are used in conjunction, each net trained to respond to one class of pattern. These nets are modified so that instead of having a gate to combine the RAMs output, the decision is left to a maximum response detector. This is shown in figure 8.7.

These modified RAM nets are known as discriminators, and the maximum response detector assigns classification to the discriminator that shows the highest response to the input pattern. A pattern is classed as "unknown" if there are equal responses from two or more discriminators, since this implies that the pattern is a member of more than one class. The type of generalisation that this arrangement demonstrates is dependent on the training data used and the pattern of connectivity of the RAM units, since this determines the sub-patterns that are encountered by each discriminator. The generalisation can also be controlled by setting a certain minimum difference between the responses of the two maximum discriminatory units, such that this difference has to be exceeded before the pattern is classed as "unknown". In this way, patterns that evoke

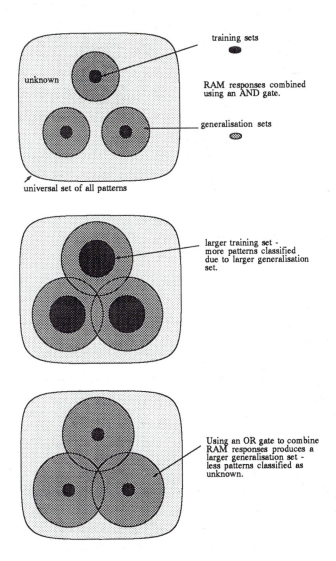

training sets

RAM responses combined
using an AND gate.

generalisation sets

unknown

universal set of all patterns

larger training set -
more patterns classified
due to larger generalisation
set.

Using an OR gate to combine
RAM responses produces a
larger generalisation set -
less patterns classified as
unknown.

Figure 8.6 Summary of the generalisation properties.

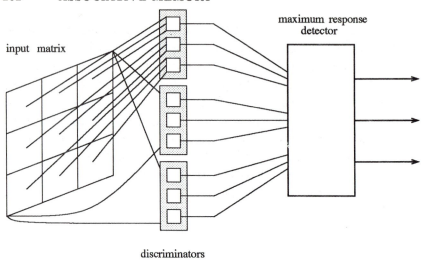

Figure 8.7 Matrix connected to three RAM units, feeding a maximum response detector.

nearly equal responses from two classes will be classified as unknown as well, therefore reducing the size of the generalisation set.

8.4   RAMS AND N-TUPLING

The RAM network was developed from a recognition process first described by Bledsoe and Browning in 1959. Known as the *n-tuple* process, it is a general form of the RAM implementation. The term "n-tuple" derives from the fact that each unit accepts n inputs as a group, or tuple. Rather than these n inputs addressing a memory location, the tuple produces an output that is dependent on the inputs, usually one bit set to 1 in 2^n possible outputs. More complicated examples are allowed in which more than one bit is set, but these are not discussed here, since their behaviour is a simple extension from the usual case.

As we can see, the tupling function outputs a unique value depending on the values of its inputs, and this output has a constant number of bits set to 1, i.e. in this case, one. This is true for tuples with a larger number of inputs, as long as there are 2^n possible output lines. This constant number of bits set to 1 is useful since it forms a sparse coding of the input. In a tuple with four input lines, for example, there may be any number of bits set to one from none to four. However, the output from the tupling function will only ever have one bit in sixteen set to 1, with the rest zero.

The n-tuple units sample parts of an input image, as shown in figure 8.8.

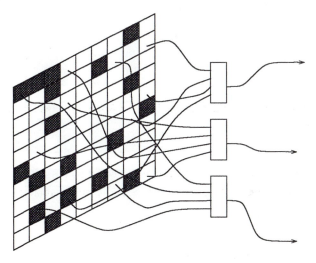

Figure 8.8 The tuples sample the image randomly.

The mapping of image bits onto the tuples is usually random, but specific mappings can be used if required. Each tuple unit "sees" a small portion of the image, and responds according to the input it receives, independently of the responses of the other tuples. If a tuple responds to a certain input pattern, this means that *any* input that has the same pattern of bits on the tuple input will provoke the same response from the tuple. This is shown in figure 8.9.

The bits that are not sampled by the tuple are free to take on either value without affecting the output of the tuple and therefore its recognition. However, these free bits are not usually ignored since another tuple may have them as its input.

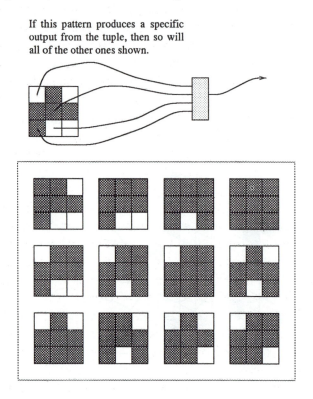

Figure 8.9 More than one local pattern can produce the same output from one tuple.

This ability to respond to patterns that have not been seen before is an essential feature of the system, and is known as generalisation.

The random mapping of the tupling, with its non-linear binary logic, means that patterns that are not linearly separable can be successfully classified, as long as the tuple sampling overlaps the pattern boundaries and so gets a different input for each of the different patterns.

8.5   WILLSHAW'S ASSOCIATIVE NET

A true associative memory is known as Willshaw's associative net. It is a distributed mapping memory, with binary inputs, "evoking" an association between the input pattern and the required output pattern. It can be visualised as a matrix of initially unlinked wires, one horizontal wire for each of the bits in the input, and a number of vertical wires, one for each bit in the output. In the teaching phase, each input example is presented along with the bit pattern with which it is to be associated. This pattern appears on the vertical wires, whilst the input appears on the horizontal wires. A "link", i.e. a weight of $+1$, is set in the memory matrix wherever an active vertical wire crosses an active horizontal wire. This process is repeated for the whole example set. This simple learning rule uses only binary links, so that once formed a link remains in place; if a new pattern requires a link in an empty position, one is formed, but if the position already has a link, then nothing is altered. The learning of patterns therefore happens in one pass through the matrix, without the need for the iterative processes that other methods require. The Willshaw net is shown in figure 8.10.

In recall, the input pattern is presented as before, and the output pattern is calculated by summing the number of links in each column that are activated by the input. These integer totals are then thresholded to recover the original binary pattern.

 We can express this mathematically as follows. Let the memory matrix $= M_{ij}$, the input vector $= A_i$, output vector B_j. Then teaching can be expressed by

$$M_{ij} = \begin{cases} 1 & A_i, B_j = 1 \\ 0 & \text{otherwise} \end{cases} \tag{8.1}$$

The recalled vector R_j is given by

$$R_j = \sum_{p=0}^{n} M_{pj}.A_p \tag{8.2}$$

This vector R is then thresholded to recover the estimate of the associated pattern B'.

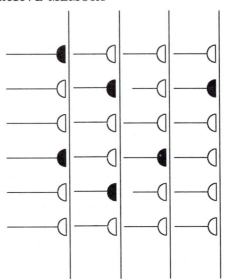

Figure 8.10 The Willshaw net: nodes in black represent weights of +1, whilst nodes in white represent zero weight or no connection.

8.5.1   Problems

There are some problems with the Willshaw net when trying to recall patterns. Whenever the output should be a 1, the net will always produce a 1, since the correct links will have been set. However, if the output should be 0, there may have been enough links set by the storage of other patterns to give a false positive output, and the net may respond with a 1 when the 0 is required. Knowing at what level to set the threshold is problematical too; if it is set too low then too many bits are set in the output pattern, but if it is too high, not enough bits are recovered. What is usually done is to choose a level that is equal to the number of bits set to one in the input pattern. For example, an input pattern with 3 bits set to one would have a threshold level of 3 as well.

Since links are set in the matrix whenever there is a 1 in the input pattern, patterns with a high proportion of 1's soon cause the net to have the vast majority of the links set and so recall becomes impossible. Such a situation is known as *saturation*, and has to be avoided in systems that strive for accurate responses.

8.6   THE ADAM SYSTEM

An improvement on the Willshaw net has been suggested by Dr. Jim Austin, and is known as the ADAM (advanced distributed associative memory) net. This incorporates the n-tupling discussed earlier as a pre-processor which samples the input and feeds the memory matrix. This matrix is in many respects the same as the Willshaw net, but is split into two parts, as shown in figure 8.11. The reason for splitting the memory into the two sections is to allow the introduction of an intermediate "class" pattern, C, which has a known number of bits set to 1. Instead of the memory storing the association A $\rightarrow$ B, it stores A $\rightarrow$ C in the first matrix, and C $\rightarrow$ B in the second. Overall, the memory has still associated A with B, but via an intermediary stage. This seems at first sight simply a little more complicated, but the introduction of the class pattern allows much more accurate recall, since the characteristics of the class pattern can be precisely determined.

The thresholding of the matrix response is done using a technique known as *n-point* thresholding rather than the standard form of setting to one all the values above or equal to a certain level, and setting the rest to zero. N-point thresholding selects the n highest values and sets those to one, returning all the remaining values to zero. This effectively gives a dynamic threshold level that is adjusted until a fixed number n of bits is recovered. This is much more successful in recalling the associated pattern than the standard static threshold method, especially as the class pattern that is being recalled has a known number of bits set to one, and so the value of n is determined. This is easiest to see with an example.

The matrix is taught with the first pattern and its associated class pattern. The second diagram shows the matrix after many other patterns have also been taught. On presentation of the first pattern, the response is shown, calculated by summing all the links that intersect with an active horizontal wire in each vertical column.

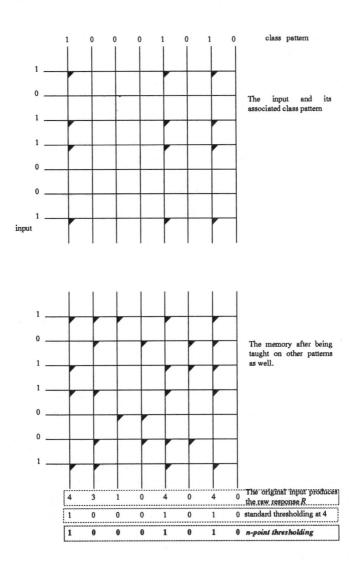

Figure 8.11 The ADAM matrix showing its appearance after teaching and the equivalence of the n-point thresholding at high response levels.

If a Willshaw threshold of 4 were used, then the response would be as shown in the second row, i.e. 10001010. Notice that this has recovered the class pattern required. If the n-point threshold is used, there were 3 bits set to one in the original class pattern, so we select the three highest values in the response, and this produces the correct output as well, i.e. 10001010. For response levels that

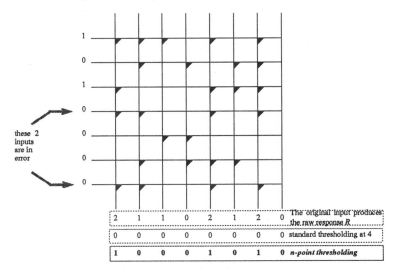

Figure 8.12 The ADAM matrix: n-point thresholding at low levels is much more successful than the fixed value approach.

are lower, however, the n-point technique is much more effective. Consider figure 8.12. The inputs to the system have two errors, resulting in a response from the matrix that is very low. Using a fixed level for thresholding does not recover any pattern at all, but the n-point technique manages to correctly produce the associated class pattern.

The ADAM system then enters a second stage, where the class pattern recalled from the matrix is passed into a second matrix, which associates this class pattern with a final output pattern. This two-stage association has a number of advantages. The class pattern acts as an intermediate stage with a known number of set bits, allowing the n-point thresholding technique to be used on noisy, incomplete, or otherwise corrupted input. This would be impossible to do if the input were associated directly with the output since there would not then be a known number of bits set to one, and so the n-point technique is inoperable. The class pattern entering the second memory is a hopefully noise-free vector that allows accurate recall in the second matrix of the final output pattern. The use of the class pattern is also storage-efficient, saving on the size of memory required. If an m by n pixel image is to be associated with an x by y output image, then $m \cdot n \cdot x \cdot y$ bits of storage are required to make the matrix. If an intermediate class pattern of a bits is used, then the storage requirements become $(m \cdot n \cdot a) + (a \cdot x \cdot y) = amn + xy$. Since $mnxy$ is much larger than $mn + xy$, space is saved. For example, if we associate a 512×512 image with itself, then m, n, x and y are all 512. Basic storage requires $512^4 = 68719476736$ bits (8589 Mbytes), but the use of a class pattern of, say, 64 bits needs storage of $64 \cdot 2 \cdot 512^2 = 33554432$ bits (4.19 Mbytes), which represents a large saving on memory space.

The use of the n-tuple preprocessing has two major advantages; it copes with non-linearly separable patterns, as we have seen, and so allows the ADAM system to resolve such problems as the XOR one. To put this in more mathematical terms, the non-linear logic in the tupling function provides a mapping that transforms any input into one that is linearly separable, given the right sampling by the tuples. It also ensures that the inputs to the memory matrix are sparsely coded, so that there are not many active lines in any input, and this helps prevent the memory matrix from becoming saturated. The whole system architecture can be seen in figure 8.13.

8.6.1   Applications

The ADAM memory was originally developed for scene analysis, although it is also used as a fast-learning network for a variety of classification problems. It has the advantage that it learns new examples with one pass through the matrix and so does not require the back-propagation of errors, or repeated iterations. However, it has no adaptive internal representation, and so cannot code higher-

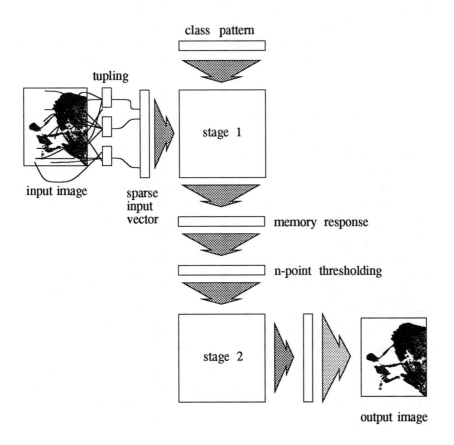

class pattern

tupling

stage 1

input image

sparse
input
vector

memory response

n-point thresholding

stage 2

output image

Figure 8.13 The ADAM system architecture.

order features of the input, unlike the multilayer perceptron. This limits its generalisation abilities, which come mainly from the tupling function. Sampling the input, the pertinent features that allow classification are not discovered explicitly but left to probability that some of them fall in regions sampled by different tuples. Since the tuple sampling is random, it is likely that important features are detected by at least some of the tuples, and generalisation occurs from these.

8.7   KANERVA'S SPARSE DISTRIBUTED MEMORY

A different implementation of associative memory was proposed by Kanerva in 1984, and can function as an autoassociative memory, a heteroassociative memory, or a sequential sequence memory. A sequential sequence memory is one in which the presentation of one pattern elicits the recall of a different pattern, which itself causes the recall of another, and so on. In a conventional memory, we have seen that data is stored by writing it into one of a number of locations each specified by a unique address, and is recalled by reading out the contents of the specified location. The addresses are represented by binary vectors, and the number of possible addresses is dependent on the length of this vector. If the length of the address vector is n, then there are 2^n unique addresses that can be accessed, and these 2^n addresses make up what is known as the *address space*. If n is large, then 2^n is very large indeed—for $n = 1000$, 2^n exceeds the number of atoms in the universe. If we wanted to consider using memories with large n, the number of physical locations quickly becomes impossibly huge and there is no way of actually implementing this amount of storage.

A method of actually implementing a memory system that was able to use large addresses was proposed by Kanerva. His approach is to randomly choose a small set of m addresses, where m is typically between a million and a billion, that are to be identified with actual storage locations. Since the value of m will be very much less than the 2^n possible addresses, these will be sparsely distributed over the address space, which is why the memory is called a "sparse distributed memory" (SDM). In order to write into this memory, both the address and the data are required, just as for a conventional memory, but the address is a bit pattern that is allowed to be any one of the 2^n possibilities. There will be a few of the m locations in memory that have their addresses close to the actual input address, and the data is written into these locations. In this context, "close" means all those addresses that lie within a Hamming distance h of the original address. In other words, if the n-bit address patterns are considered to lie in an n-dimensional space, then all the selected patterns will correspond to all those physical locations that have addresses within a hypersphere of radius h centred on the actual input address. This is shown in figure 8.14.

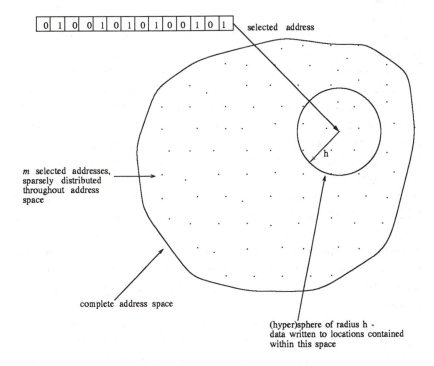

Figure 8.14 Diagrammatic form of the SDM, demonstrating the Hamming hypersphere containing the selected addresses.

Instead of overwriting the contents of the previous value stored in any selected location, the data is added in, since there may be occasions when we have to write two or more sets of data into the same location. This will occur if the hyperspheres chosen by different input addresses are sufficiently close together to overlap, causing the selection of the same address by different inputs. This means that each location of the SDM consists of a set of n counters. For the system to be effective, the vectors are considered to consist of bipolar $(+1, -1)$ values rather than binary $(1, 0)$ values, since the 0's in binary vectors are ignored when added, but the -1's in bipolar vectors are not: $1 + 0 = 1$, whereas $1 + (-1) = 0$. For recall in the SDM, all the selected locations that lie within the Hamming distance of the input address are read, and the values in each of the n counters are added in parallel to yield n sums. Each of these sums is then thresholded at zero, with a $+1$ output if the sum is greater than zero, -1 if it is less than zero, and the value remaining unchanged if it happens to equal zero. This threshold process is usually able to separate out the required pattern from the corrupting overlap of other similar patterns, if not too many patterns have been stored.

The advantage of such a system is that it enables large address patterns to be associated with physical storage locations, and so complex inputs that are represented as large bit patterns can act as the address for storage. This means that the SDM acts as a *content-addressable* memory. What is more, since the actual storage locations that are accessed lie within a certain Hamming distance of the address provided, most of the locations accessed will be the same if an input address has a small number of incorrect bits. This ensures that recall is still possible even if the addressing pattern contains a few errors, and so the SDM can act as a true associative system. In other words, a slightly corrupted input pattern should still lie within the hypersphere of the actual address centre, and so the data recovered will be what was originally stored. The memory functions as a heteroassociative one if the data vector stored is of a different size to the accessing data address, and as an autoassociative memory when the data stored is actually the address. The principles for autoassociative recall hold for the sequential sequence case, only the recovered output is taken as the new input to generate the next pattern.

8.8   BIDIRECTIONAL ASSOCIATIVE MEMORIES

Bidirectional associative memories (BAMs) were proposed in 1988 by Kosko, and they can be seen as a two-layer non-linear feedback network, as shown in figure 8.15. Patterns sweep from one neuron layer to the next, and then back again, slowly relaxing into a stable state that represents the network's association of the two patterns.

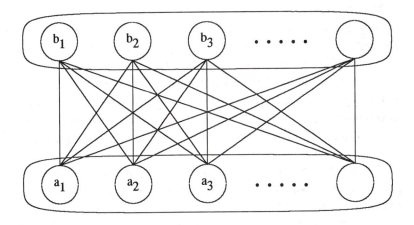

Figure 8.15 The BAM seen as a two-layer network.

The weights in the forward pass can be represented as a connection matrix M, whilst those in the backward pass are given by the transpose of this matrix, denoted M^T. The use of the connection matrix's transpose makes the BAM interesting, since this distinguishes it from other systems which use a different matrix of connections for the backward pass.

The BAM stores pairs of patterns A_i, B_i, and is autoassociative if $B_i = A_i$ and heteroassociative if B_i is different from A_i. In a standard heteroassociative memory such as the Willshaw net discussed earlier, A is presented to M, then thresholded to produce output B

that is hopefully closer to the stored pattern B_i than to all the other patterns B_j, if A was closer to A_i. However, we have seen that this assumption is not always valid, and we would like a procedure that would allow us to increase the accuracy of the final recall. The BAM achieves this by passing the output B back through the system to produce a new value, A', which should be closer to the stored pattern A_i than was the original pattern A. This new value is passed forwards again, producing a better estimate B', and the process repeats until it settles down to a steady resonance between the stored patterns A_i and B_i. The advantage of using the transpose of the matrix M^T is that it requires no additional information, and this information is locally available to each node. Kosko has proved that the BAM converges to a fixed pair of stored patterns by extending Hopfield's argument, and demonstrates that the Hopfield case of autoassociation is simply a specialised case of the BAM, when $B = A$. In other words, the sequence $A \to M \to B$, followed by $B \to M^T \to A'$, which continues, producing a series of approximations $(A, B), (A', B'), (A'', B''), \ldots$ will converge to a steady resonant state that reverberates between the fixed pairs (A_f, B_f). Having proved that any matrix M is bidirectionally stable in this way, he goes on to show that patterns cannot only be recalled from a fixed matrix M, as in the Hopfield net, but that if small changes are made to M in accordance with a Hebbian learning rule, it will *learn* to associate two patterns. In this case, as the patterns oscillate back and forth, pattern information is allowed to seep into the weights, resulting in the learning of an association between the two patterns.

8.9   CONCLUSION

We have discussed the main principles of associative memory, as well as focussing on the major approaches currently in use. The distinction between these forms of memories and neural networks is a hazy one, since each can play the role of the other. The approaches to associative memory tend to offer advantages in the speed of storage of patterns, but are unable to perform the complex data representation tasks in the same way that multilayer perceptrons can. This means that practical decisions as to the suitability of one method over another have to be carefully considered.

 Summary

- Two stored patterns are autoassociative if they are the same, and heteroassociative if they are different.
- Sequential access memories retrieve one pattern after another via autoassociation.
- Content-addressable memory (CAM) is accessed via knowledge of its contents, not its address.
- Hash coding implements CAM.
- Random access memories used to implement associative memory.
- n-tupling takes many small samples from an image. Produces sparsely coded output.
- Willshaw net provides associative memory. Matrix of binary connections set to 1 whenever both active. Requires thresholding to recover pattern.
- ADAM provides more effective thresholding and storage than Willshaw net by using n-tuple preprocessing and n-point thresholding to recover intermediate pattern with known number of bits.
- Sparse distributed memory implemented by selecting a few physical locations to represent many similar addresses.
- Bidirectional associative memory resonates two patterns via a matrix and its transpose.

FURTHER READING

1. *Self Organisation and Associative Memory*, third edition. T. Kohonen. Springer-Verlag, 1990.

2. Guide to pattern recognition using random-access memories. I. Aleksander & T. J. Stonham. In *Computers and Digital Techniques*. Volume 2, number 1, 1979. A review of RAM pattern recognition systems.

3. ADAM: A Distributed Associative Memory for Scene Analysis. J. Austin. In *Proc. First Int. Conf. on Neural Networks, IEEE*. Eds. M. Caudhill & C. Butler. Volume 4, 1987.

4. *Sparse Distributed Memory*. Pentti Kanerva. MIT Bradford Press, 1988. Kanerva's own book on his sparse memory system.

5. Bidirectional Associative Memories. Bart Kosko. In *IEEE Transactions on Systems, Man, and Cybernetics*. Volume 18, number 1. January/February 1988. An interesting paper, it produces the BAM and gives a neural network interpretation of it.

9

Into the Looking Glass

9.1 OVERVIEW

The purpose of this chapter to to look ahead to the future of neural computing. There are two major areas of implementation that are developing rapidly: the hardware neural network chips, and the optical computing field. The mathematical techniques used in the analysis of networks are also becoming more diverse, and improvements in understanding can be expected from developments in the theoretical areas of the subject. The interchange of ideas across the boundaries of scientific disciplines means that it is practically impossible to predict what the future has in store, but the two areas of hardware realisation both have great potential. It is not the purpose of this chapter to be comprehensive in scope and description, but to paint the broad outlines of future developments.

9.2 HARDWARE AND SOFTWARE IMPLEMENTATIONS

The majority of the networks that we have discussed exist as software simulations only, barring the optical Hopfield and RAM associative memory networks. By this we mean it is not possible to buy integrated circuits that contain an artificial neural network. The results and applications that have been quoted in this book all stem from software simulations on standard computer hardware from IBM-PCs to high-performance parallel machines. The reason that we have included the algorithms in each chapter is primarily so that the interested reader can actually code them into programs. We recommend that you consider doing this because it provides a very useful insight into the workings of the learning methods for these algorithms.

If, however, you are not in a position to code the algorithms yourself, then you may wish to consider one of the numerous software packages available that simulate most of the major neural paradigms. These are available from such companies as Nestor, Hecht-Nielson, Science Application International Corporation (SAIC) and Neuralware, to name but a few. New software products are regularly arriving on the market, with prices ranging from anywhere between twenty and ten thousand pounds. It might be worth adding a note of caution about the computing resources that are required to run typical software simulations of neural networks. One common feature that all the various algorithms share is a significant amount of "number-crunching"—any network of practical dimensions will place heavy demands on the processing power required during training. The main mathematical processing is the multiply and add for the weights of each node in the network. For methods such as backpropagation, there is also the error derivative for gradient descent learning. Any computer that is to run simulations of neural networks ideally requires a large amount of storage memory (to deal with the large vector matrices) and a fast microprocessor. Typically, software simulation of neural networks is performed using computers with add-on accelerator boards (or co-processor boards) that have high-performance processors on them, capable of very fast multiply and add operations. On slow computers, without these boards, it is not unreasonable to expect training times of several hours or even days for some applications.

This leads us on to think about hardware for neural networks. Although we have already said that there are no commercially available neural network integrated circuits, there are actually several large electronic device companies about to release such products. There are many practical difficulties in implementing a neural network at chip level—the most obvious of which is that neural networks, by nature, are complex adaptive systems. It is very difficult to implement adaptive weights in integrated circuit technology. Three approaches are currently taken: analogue, digital and fixed weight. Analogue techniques for creating modifiable weight connections include variable resistors, FET gate voltage control and capacitive storage methods. However, the major drawback of most of these methods is that they require large amounts of silicon space, resulting in only a very low density of neural nodes available on a chip. Digital techniques use addressable registers to store and modify the weights. This technique is useful but is again limited by the space required for multiply and add units on the silicon. The third alternative avoids the problem of modifying weight values by only allowing the value of the weights to be set once. The idea behind this is to learn the correct weight matrix, in a simulation environment, and then load this into the chip permanently.

All the methods also suffer from the other drawback of neural networks, namely high interconnectivity. It is both costly and difficult to design integrated circuits wth complex data pathways between the layers of nodes, and even when it is achieved it invariably means that the topology of the neural network is fixed. These restrictions mean that in many application environments integrated circuit technology is just not suitable. VLSI technology is advancing at a remarkable rate, however, and these implementation difficulties will not hinder progress of neural network chips for too long. One technology that may provide some answers, particularly to the interconnectivity problem, is optical computing.

9.3   OPTICAL COMPUTING

9.3.1   Introduction

The purpose of this section is to give a very brief overview of the developments taking place in optical computing, with particular reference to the effects these may have on the artificial neural systems of the future. A comprehensive review is outside the scope of this book; this is simply meant to sketch the broadest of outlines.

9.3.2   What is Optical Computing?

In order to compute we need to transport data from place to place, connect components together, store data, and be able to switch on and off. The electronic equivalents of these functions are wires or conducting pathways on silicon, electrical junctions, memory, and the transistor.

Optical computing uses light to transport information instead of electrical signals. This approach holds two major advantages for computation in general and artificial neural networks in particular. The first is in the inherently high speeds achievable—data can flow at the speed of light, and optical switches can go much faster than electronic ones. However, for neural computing in particular, the more important reason is due to the fact that one beam of light can cross another and emerge completely unaffected by its encounter, whereas two electrical wires cannot. This opens up the potential for massive interconnectivity within a small space.

A simple lens can be thought of as a powerful interconnection device. The image that it forms is a collection of rays of light reflected from the object, and the lens effectively connects millions of these rays from the object to the image, as shown in figure 9.1. These light

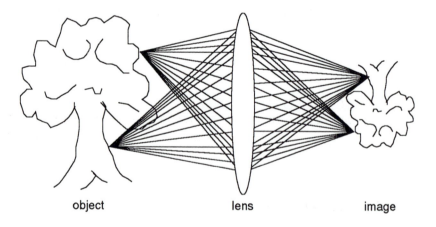

object   lens   image

Figure 9.1 A lens offers immense interconnectivity.

rays can come close together and cross without affecting the data carried in either one, which is why the image can be formed. Such huge connection densities are impossible to achieve with electrical circuits even if they are routed on silicon, since each path must be a certain distance from its neighbours to avoid interference.

Storage in optical systems is accomplished using holograms. The physical principles underlying the hologram are not relevant to this book—suffice to say that holograms are a sort of three-dimensional photograph, containing enough information to reconstruct an image of a solid object. Holograms can also be used as switches by directing light that falls on to them in different directions dependent on the initial angle of approach of the beam. The amount of information that can be stored in a hologram is huge, since a single one can hold many images.

Optical switches can also be made. One approach is to affect a crystal structure with an electric or magnetic field, which alters its optical properties, and so affects incoming light differently. These work at speeds of around 10^{-10} seconds, compared to the best transistor switching times of down to 10^{-12} seconds. Other switching devices use non-linear crystals that alter the amount of light that they transmit depending on the intensity of the incoming beam. The best optical switches are currently switching at speeds up to 10^{-14} seconds, which gives them the speed edge over electronic ones.

9.4   OPTICAL COMPUTING AND NEURAL NETWORKS

Optical influences on neural networks fall into one of two areas, either in implementing parallel matrix multipliers or in holographic pattern recognisers.

9.4.1   Matrix Multiplication

Many of the operations in networks require the evaluation of a set of inputs multiplied by some weight matrix, and this process can be implemented in an optical system. If the weight matrix is W_{ij}, then the weighted sum of the inputs X_i to a unit j is given by

$$Y_j = \sum_i W_{ij} X_i.$$

We can see that the i-th input is only of interest to the elements in the i-th row in determining the result. The inputs are represented by a beam of light and are spread out to span the rows of the grid. Each element of the grid contains a piece of photographic film whose transmittance is proportional to the value of the weight in the matrix. A photodetector receives its input from a lens that gathers all the light that emerges from one column of the grid, and the intensity of the light it receives is a sum of values that depend on the product of the intensity of the input signal and the transmittance of the "weight" through which it has passed. This is shown in figure 9.2.

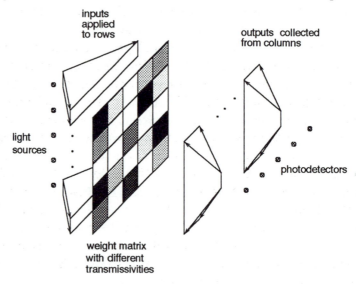

Figure 9.2 An optical matrix multiplier.

The summing operation can occur in parallel, and the speed of the system is independent of the size of the weight array, and so can be scaled up without becoming any slower. This opens the way for very large networks with correspondingly large weight matrices. Currently the weight matrix has to be altered by substituting a different mask, so automatic learning is not possible, but research is in progress to investigate the use of liquid crystal cells which could have a variable density.

9.4.2   Holographic Pattern Recognition

Holographic pattern recognisers are essentially resonant systems; a typical example, due to Abu-Mostafa and Psaltis, (Scientific American, March 1987), is shown in figure 9.3.

The key to the operation of the system is the threshold device. This is a non-linear reflector, which reflects most strongly from its front surface the pattern that appears brightest on its back. The input is passed to a beam splitter which sends one copy of the input on the front of the threshold device, and passes another to a hologram. This hologram contains several stored images that represent the patterns that the system is to recognise. The input pattern is passed through this hologram, which correlates these patterns and the input. The correlations are a measure of the similarity between the patterns, and the pattern that is the most similar is the brightest. This is passed through a pinhole which separates the images, and via a mirror and lens through another hologram like the first. This correlates the new images, and passes the results to the rear of the threshold device. The back of the threshold device therefore receives a set of images corresponding to the stored images in the system. The brightest one of these will be the one that the original image was most similar to, and this means that this pattern will be most strongly reflected from the front of the device. This new enhanced pattern will then pass round the loop for further enhancement, and the system will quickly settle into a state in which the pattern most like the input pattern goes round and round the loop until stopped. The speed that the system relaxes into this steady state is impressive, and it is capable of recovering an image when only a very small proportion of the original is presented.

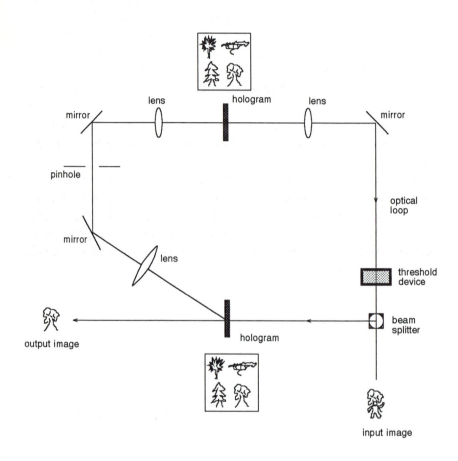

Figure 9.3 A holographic pattern recognition system.

9.4.3   Conclusion

The question arises as to why all neural networks are not currently built optically; the answer is that the current technology is unable to provide images of a reasonable quality. The holographic recognition system is bulky and difficult to align correctly, but is likely to improve its performance as further research is done. Such systems are pushing the barriers of technology to the limits; however, advances will be made given time and money. Another difficulty arises if the optical system is to be part of a larger electronic one, since there is then the need for an optical–electronic interface between the two. Whilst it is difficult to provide a good interface between the two types of systems, Demetri Psaltis and his co-workers have prepared holographic memories using an electronically-addressed array of lights as the input. Optical systems offer intrinsic parallelism, the potential for massive interconnectivity within a small volume, and computation speeds substantially faster than electronic approaches. The time will come when such esoteric systems will become much more commonplace.

Index